D0405278

More praise for *The Eureka E...*

"This cornucopia of brain-teasers tests your mettle, sharpens your skills, and illuminates the mysteries of human problem-solving." —Howard Gardner,
Hobbs Professor of Cognition and Education,
Harvard University, author of *The Disciplined Mind*

"All too often, [Perkins] notes, humankind and nature both come up against unreasonable problems that call for breakthroughs. And that, as he repeatedly shows us in [*The Eureka Effect*], is where the fun is."
—*San Jose Mercury News*

"Perkins' style is engaging—not for eggheads only—and the brainteasers are entertaining and surprisingly fresh."
—*Detroit Free Press*

"Perkins grounds the ineffable processes of cognitive thought in playful yet crystalline prose."
—*The New Republic*

"For scientists, for whom creativity is paramount, [this book] offers much in terms of how we can structure, develop and interpret our research."
—*Nature Cell Biology*

THE EUREKA EFFECT

The Art and Logic of Breakthrough Thinking

DAVID PERKINS

W·W·NORTON & COMPANY
New York · London

Copyright © 2000 by David Perkins

All rights reserved
Printed in the United States of America

First published as a Norton paperback 2001

For information about permission to reproduce selections from this book,
write to Permissions, W. W. Norton & Company, Inc., 500 Fifth Avenue,
New York, NY 10110

The text of this book is composed in Optima and Galliard with the display
set in Galliard.
Composition by Thomas Ernst.
Manufacturing by Haddon Craftsmen Inc.
Book design by BTDnyc.

Library of Congress Cataloging-in-Publication Data

Perkins, David N.
    Archimedes' bathtub : the art and logic of breakthrough thinking / by David
Perkins.
        p.   cm.
    Includes bibliographical references and index.
ISBN 0-393-04795-4
    1. Thought and thinking. 2. Inspiration. 3. Creative thinking. I. Title.

BF441 .P47 2000
153.4—dc21   00-020012

ISBN 0-393-32255-6 pbk.

Originally published under the title *Archimedes' Bathtub*

W. W. Norton & Company, Inc.
500 Fifth Avenue, New York, N.Y. 10110
www.wwnorton.com

W. W. Norton & Company Ltd.
Castle House, 75/76 Wells Street, London W1T 3QT

1 2 3 4 5 6 7 8 9 0

To my sons Theodore and Thomas Perkins,

who talked over with me several of the ideas in this book, solved many of the illustrative puzzles within, and themselves invented a couple lurking in wait for the reader.

Contents _____

Acknowledgments ⸺

This is one of those books that, like a painting by Monet, dips into many colors to put the picture together. Rounding up the relevant sources and extracting key information have been a substantial task not only for me but for three individuals for whose assiduous and intelligent help I am most grateful: Dorothy MacGillivray, my administrative assistant for several years; her assistant Cynthia Rogers; and my current administrative assistant, Lisa Frontado.

A particular pleasure of writing this book was to pursue its theories through informal experiments with insight puzzles, trying them out on family, friends, and sometimes my students at the Harvard Graduate School of Education. Frequent respondents as well as conversationalists about some of these ideas were my two sons, Theodore and Thomas Perkins, to whom this book is dedicated. Two other especially helpful individuals as well as longtime colleagues were Abigail Lipson and Beatriz Capdevielle.

I would also like to acknowledge generally those ingenious folk around the world who have made up insight puzzles and written them down or passed them along orally to others. When I have a specific written source that provided a puzzle, I have cited it. Also, a few of those used here were devised by Ted and Tom or myself. But often, these puzzles have come to me from others through incidental conversations, sometimes essentially the same puzzle in different forms. Like jokes, they drift around on the seas of casual conversation. It is often difficult to assign a home port.

Special appreciation goes to my longtime colleague Shari Tishman, who read the draft and recommended a number of large and small adjustments that have strengthened the book. Finally, I would like to thank two individuals professionally committed to the world of publishing for their good counsel in bringing this project to fruition: Faith Hamlin, my agent, and Ed Barber, my editor at Norton. The pages that follow are certainly much the better for their ministrations.

Part 1

The Idea of Breakthrough Thinking

In which we survey how breakthrough thinking has shaped science, art, engineering, and more; find a similar process in biological evolution; analyze the deep structure of how breakthroughs occur; and practice breakthrough thinking with some puzzle problems.

1

Thinking like Leonardo

Thinking Takes Flight

Leonardo da Vinci was wrong. But he was *insightfully* wrong. He came to a mistaken idea about flight, but the pattern of thinking behind the idea was exemplary.

The fifteenth-century Italian who combined art and science so brilliantly has a stellar reputation even today for a roving and flexible mind. From the *Mona Lisa* to the design of engines of war, from the *Last Supper* to anatomical dissections that disclosed the subtle mechanisms of the human body, Leonardo sought to understand the world around him, express visions of it, and translate his conceptions into practical devices. Indeed, it could be said that Leonardo attempted too much. He became notorious for starting projects and not finishing them. Busy fellow that he was, inevitably he would make mistakes.

Leonardo attained a number of creative breakthroughs about flight. He observed birds carefully, analyzed their mechanisms, and formulated principles. He recognized that birds achieved flight not just by flapping but by riding up ramps of air: "Birds which rise on the wind in circles hold their wings very high, so that the wind may serve as a wedge to raise them up." Working from such observations, he sketched several flying devices for human beings. One of them was a sort of helicopter, with a giant screw of radius 8 *braccia,* or about 14 feet. Leonardo wrote: "I find that if this instrument made with a screw be well made, that is to say made of linen of which the pores are stopped up with starch, and be turned swiftly the said screw will make its spiral in the air and it will rise high."

Leonardo's insight made a connection between two very different things. He saw a relationship between screws and the challenge of flight. A propeller amounts to an air screw, holding on to air much as a wood screw holds on to wood, albeit less firmly.

Leonardo certainly never got his idea to work on a human scale. It was quite unworkable. The screw shape he envisioned would have to carry too much weight to get airborne with any power human muscles could provide. Basic issues of physics stand in the way: Flight is easier for small creatures like birds or insects—or small vehicles like toy helicopters—than large ones, which require proportionately much greater energy sources. It was not until the beginning of the twentieth century that Wilbur and Orville

Wright finally solved the many problems of heavier-than-air flight. They did it by thinking like Leonardo da Vinci.

The Wright brothers also took inspiration from the flight of birds, but they added their own distinctive breakthroughs. One key realization concerned the propeller. Initially, the Wright brothers assumed that they could base the design of their aeronautical propeller on some theory of propeller design found in marine engineering textbooks. They soon discovered that there was no such theory. Relying on their own ingenuity, the Wright brothers reached for an analogy, as had Leonardo da Vinci, but came up with a somewhat different one. A propeller should be thought of not as an air screw but a rotary wing. Just as the wings of the plane would give it upward lift, the "wings" of the propeller would give it a forward pull. The wing-propeller analogy allowed the Wright brothers to apply their technical analysis of wing design to propeller design as well. The development of the propeller was one of the last major breakthroughs necessary for the invention of powered flight.

The most striking thing about Leonardo da Vinci's thinking, and the Wright brothers', was not the conclusions they reached but the paths they took. Both looked to analogies to reframe a puzzling problem and find an unexpected solution. Leonardo's attempt was so far ahead of his time that he lacked key pieces of the puzzle. The Wright brothers, supported by a plethora of advances in science and technology, made it work. Beyond the particulars of flight or any other technological invention, both can be admired for *breakthrough thinking*, a kind of thinking that has helped much of the world's population toward exceptional levels of comfort, health, and understanding.

This book asks and tries to answer a simple question: How does breakthrough thinking work? Notice that the focus is not just on creativity in general, but breakthrough thinking specifically, the sort of creativity that makes a decisive break with the past. What kinds of quirky problems, exotic mental processes, clever thinking strategies, and other ingredients go into it and determine its nature, as flour, eggs, and so on make a cake a cake? In the spirit of the theme itself, the answer will turn out to be rather different than one might expect.

Eureka!

Any breakthrough worth its salt is worth an exclamation. Most of us would probably say "Aha!" but we might say "Eureka!" *Eureka* is a word from ancient Greek meaning "I have found it!" It's curious that a term more than two thousand years old should be in anyone's vocabulary today. The reason lies in another story of breakthrough thinking, the one that gives this book its name, the discovery by Greek philosopher and scientist Archimedes, of the principle of the displacement of water. Although probably apocryphal, it is one of those tales that ought to be true, so well does it illustrate the phenomenon.

The occasion was a royal command performance. Hiero II had become the new ruler of Syracuse. He thought himself favored by the gods and arranged for a golden crown to be made, aiming to dedicate it to them in thanks. Hiero provided the gold and soon received the crown from the craftsmen. However, he heard that the craftsmen might have stolen some of the gold. The crown weighed the same as the gold Hiero had supplied, but perhaps the craftsmen had substituted less-valuable silver for some of the gold— not enough to change the color but enough to make a tidy profit. Not about to be cheated, Hiero asked Archimedes to determine whether the crown contained all the gold. Archimedes knew that silver was not as dense as gold. If the craftsmen had substituted an equal weight of silver for gold, that would make up more volume in silver than the gold it displaced. The crown would be slightly larger than it should be. Even so, the problem was not easy. How could Archimedes determine the volume of the crown, an object with a highly irregular shape, to check whether it was suspiciously large?

With the puzzle much on his mind, Archimedes went to the public baths. As he settled into a tub, he noticed the water overflowing the sides. The deeper he settled, the more water spilled over. In a flash, Archimedes discovered his answer: His body displaced an equal volume of water. Likewise, by immersing the crown in water, Archimedes could determine its volume and compare that with the volume of an equal weight of gold. Legend says that Archimedes leaped from the bath and ran through the streets of Syracuse naked shouting "Eureka!"—I have found it. In his use of analogy, Archimedes was thinking like Leonardo. Or, since Archimedes came first, we might better say that Leonardo was thinking like Archimedes.

How Breakthrough Thinking Built Our World

As I sit and write, that quintessential invention the lightbulb illuminates my work area, thanks to the breakthrough thinking of Thomas Edison around 1880. I put down words with the help of a desktop computer, technological descendant of simple transistors developed by Bell Laboratories in 1948 that have since mutated to fill a thousand technical nooks and crannies. And if all this seems too contemporary, I sit in a simple chair.

Consider for a moment the origins of that chair. Chairlike forms rarely occur in nature. Our forebears would have found the seat in a handy rock or fallen log. They would have found the back in the trunk of a tree or the wall of a cave. They would have made do, alternating between backless seats and seatless backs, until someone did some small-scale breakthrough thinking. The chair doesn't seem that difficult. Nonetheless, a chair is not a completely obvious construction. It is not given by nature, handed over for human use as transparently as a fallen log across a creek makes a bridge.

Not incremental but *transformative* invention is the theme here. Cuts and wounds have been minor, unproblematic events for several decades because of antibiotics—and who would have thought that living organisms like molds would generate agents that could stop bacterial infections? Most languages of the world have phonetic alphabets, with characters standing for sounds rather than the more obvious words or ideas, a transformative solution. For hundreds of years, artists painted buildings with roof and wall lines converging in a vaguely perspective-like way. But it took Filippo Brunelleschi and others, thinking hard about the implications of optics and the paths of light rays, to equip the early Renaissance with true perspective. Laces hold the hood of my winter parka shut, not such a subtle idea. But there are much less obvious devices as buttons, a zipper, and Velcro. My jacket is practically a history of fasteners.

Gadgets, philosophies, forms of government, systems of symbols, industrial processes, and many other advances emerge not through steady improvements but significant leaps beyond whatever served before. This is not to say that more incremental thinking hasn't made a creative contribution. A good deal of very worthwhile creative thinking yields not breakthroughs but rather valuable and somewhat novel ideas and products within an established framework. But it *is* to say that when we look at true breakthrough thinking, we are examining something special.

If breakthrough thinking built our world, or at least a good deal of it, what can be said about the process of breakthrough thinking? Simply on the basis of historical examples, it's useful to set out an initial sketch. In Archimedes' adventure at the baths, and in many similar episodes, one can discern a fivefold structure that goes something like this:

1. *Long search.* Breakthrough thinking characteristically requires a long search. Archimedes struggled with Hiero's problem. Leonardo da Vinci fussed endlessly with flight, and the Wright brothers dedicated years to their quest.

2. *Little apparent progress.* A typical breakthrough arrives after little or no apparent progress. Archimedes struggled with Hiero's problem. The Wright brothers wasted time on nautically inspired versions of the airplane propeller.

3. *Precipitating event.* The typical breakthrough begins with a precipitating event. Sometimes external circumstances cue this moment: the water overflowing in Archimedes' bath. Sometimes a mental event is sufficient: finding the right point of view, such as thinking of a propeller as a screw or a wing.

4. *Cognitive snap.* The breakthrough comes rapidly, a kind of falling into place, a cognitive snap. Not much time separates the precipitating event from a solution, even if

details remain to be checked. This is Archimedes' classic
"Eureka!" The time might be a fraction of a second, several minutes, or sometimes more, but the time is characteristically short compared with the time leading up to
the breakthrough. In contrast, working out the ramifications of the basic insight may take some time. The
Wright brothers labored and argued for months on the
technical problems of translating their propeller-as-wing
insight into a well-engineered propeller.

5. *Transformation.* The breakthrough transforms one's
mental or physical world in a generative way. Archimedes
certainly did not have anything like the principle of displacement of water in mind prior to his bath. Propellers
were not screws or wings until Leonardo and the Wright
brothers made them so. Ideas like the principle of displacement of water and technologies like flight also have
a profound practical impact, altering the way we human
beings get things done.

Such a sketch stands a long way from a theory of breakthrough thinking. The five steps reveal very little about
what goes on in the mind that allows it to cross the great
divide between a clutter of unrelated precedents and a new
synthesis. Nonetheless, this sketch will help to keep in
focus some important features of the process with an eye
to explaining them later.

Malthus Kicks a Field Goal Twice

The precipitating event and cognitive snap are a little like
kicking a field goal in football: First comes the kick, and a

few seconds later the football flies between the goalposts. In the examples so far, analogy has figured prominently in these mental field goals—Leonardo's propeller as screw, the Wright brothers' propeller as wing, Archimedes' body displacing water as crown. But is analogy the only recourse? An expanded picture of the possibilities comes from one of the premier scientific discoveries of all time, the development of the theory of evolution.

In September of 1838, Charles Darwin had been fifteen months home from his famous Voyage on the H.M.S. *Beagle*. During the voyage, he had found compelling evidence for the reality of evolution in his investigations of bird species on the Galapagos Islands. Convinced of the reality of evolution, Darwin cast about for a mechanism. If evolution was a fact, what made it happen?

Darwin filled his notebooks with speculations about how evolution might occur, some of them rather bizarre. The crucial breakthrough came during a moment away from the problem. Darwin reported that he was reading Malthus's famous *Essay on the Principle of Population* for entertainment. The essay explored how the unchecked exponential expansion of human populations leads to disaster due to overpopulation. Pondering Malthus's thesis, Darwin realized that the fittest organisms would survive and pass on their traits to their offspring. This could provide a mechanism for evolution. In a few moments, Darwin achieved an insight that resolved a puzzle he had been working on for months, an insight that eventually changed the thinking of humankind about itself.

Interestingly, Darwin did not fully recognize the significance of his discovery immediately. Psychologist Howard Gruber, in a close reading of Darwin's notebooks, argues that he did not experience this discovery as an epiphany.

Darwin simply wrote the idea of natural selection down in his notebook as he had dozens of other speculations about a mechanism for evolution. He did not turn back to it for some days. Only then did he reexamine it, start to recognize its fertility, and draw out its implications. One might say that Darwin experienced a slow snap rather than a quick one. Some discoveries require a while for their potential to be recognized—hours, days, or even more. Even so, the time in question is still much less than that invested in the overall search. The snap is a snap relative to the period that comes before it.

Darwin himself did not give a fine-grained account of how he made the critical connection. He simply wrote in his biography that in 1838, upon reading Malthus's essay with its dire predictions of overpopulation, "it at once struck me that under these circumstances favorable variations would tend to be preserved, and unfavorable ones to be destroyed." However, fearing a backlash from those committed to the biblical story of creation, Darwin delayed publishing his idea for twenty years while he gathered evidence for what he knew would be a controversial stance. In 1858, Alfred Russel Wallace made the same discovery. Strangely, he also made it in response to reading Malthus, who therefore kicked this field goal twice. Wallace had read Malthus's essay some years before, but it happened to come to mind, and he pondered the implications of what Malthus said. Wallace had more to say than Darwin about the link between Malthus's ideas and natural selection. Streamlined a bit, it amounts to the following:

1. Malthus wrote of humans, not animals. He emphasized how disease, accidents, war, famine, and like factors kept the population of the "savage races" down.

2. Wallace saw that this applied to animals too, an easy analogy.

3. Animals breed more rapidly than humans, so this meant a high death rate.

4. Pondering this, Wallace asked himself, "Why do some die and some live?"

5. The answer seemed obvious: "on the whole, the best fitted live. From the effects of disease the most healthy escaped; from enemies, the strongest, the swiftest, or the most cunning; from famine, the best hunters or those with the best digestion."

6. "Then," Wallace concluded, "it suddenly flashed upon me that this self-acting process would necessarily *improve the race,* because in every generation the inferior would inevitably be killed off and the superior would remain—that is, the *fittest would survive.*"

There is no way to know whether Wallace's autobiographical account describes his course of thought accurately. Nor does his report mean that he reasoned it out painstakingly step by step. After all, he uses phrases like "it suddenly flashed on me"—a cognitive snap. But these uncertainties aside, Wallace's expression of the connection is illuminating in three ways. First of all, analogy, although present, does not appear to be the mainspring. The analogical part is very easy: Human beings are, after all, biological organisms. Second, logical connections play a role. Wallace does not describe some leap into the void that landed in a happy place, but a chain of reasonable relationships. Third, asking a crucial question figures centrally. "Why do some die and some live?" Another person might not have asked that question. So the repertoire of moves behind breakthrough thinking expands to include not only analogy but logical extrapolation and asking the right question.

Ninety-Nine Percent Perspiration

If the precipitating event and cognitive snap are key moments in breakthrough thinking, the long search with little progress also makes its contribution. Sometimes the long search may acquaint one deeply with the problem at hand and put one in a position to recognize the precipitating event when it happens. As Louis Pasteur was noted for saying, "Chance favors the prepared mind." But sometimes it's much simpler than that: To find something, you need to move around a lot in the right neighborhood.

It was in this spirit that Thomas Edison called invention 1 percent inspiration and 99 percent perspiration. He was certainly underplaying his ingenuity, but he also knew what he was talking about. He elevated to an art the process of scanning large numbers of possibilities systematically. He even had a name for the process: a draghunt.

A famous example was Edison's search for a suitable filament for the lightbulb. The challenge was a tricky one. Efficient generation of light required raising the filament's temperature close to its melting point. Unfortunately, such a process tended to cause filament materials to expand, crack, and fuse. Edison did some preliminary tests with carbonized paper—paper partially burned to reduce it to the carbon it contained, which would conduct electricity. However, Edison found that it burned out very quickly, even in the vacuum of a lightbulb.

Previous work by others attempting to develop a light-bulb had used a promising metal. Platinum had a very high resistance and could be heated to 5,000 degrees without burning out. These two properties made the platinum filament highly suitable for the vast lighting system that Edison envisioned. He began experiments with platinum, but soon tested a variety of other metals as well. Palladium cracked and even bubbled upon heating. Gold could not reach incandescence. Other discouraging results were found with ruthenium, iridium, and rhodium. Edison did obtain some promising early results with nickel. However, he soon discovered that nickel oxidized too quickly to be suitable for the filament. These results led Edison back to platinum.

However, economic realities forced Edison to abandon the platinum filament. He realized that the necessary quantities of platinum for mass production of lightbulbs would be both difficult to find and expensive. This led Edison back to carbon, which was more accessible and economically feasible. In fact, Edison's own laboratory, Menlo Park, already had a vast amount of carbon for the chalk-drum telephone receiver it was manufacturing. Carbon also seemed a promising direction because any number of carbon-based materials can be carbonized. Edison and his colleagues tried a startling range of materials, including fish line, cardboard, and soft paper, carbonizing them and testing them as filaments. The cotton thread worked the best, except that it proved too fragile. This led Edison to test other materials such as wood shavings, flax, and even coconut hair. For a brief period Edison settled on carbonized paper as the material for the filament, a substance he had tested earlier and rejected. However, after receiving a letter from scientist Simon Newcomb advising him that light output would be greater with a more homogeneous and solid form of carbon, Edison continued his search. After trying many different plant fibers, in the summer of 1880 Edison settled on carbonized bamboo fiber.

Later on, Edison's final choice fell to another competitor in the marketplace. The tungsten filament lamp was first developed by Austrians Alexander Just and Franz Hanaman and first produced in the United States in 1907. However, the filament was not really successful until 1908 when William D. Coolidge developed ductile tungsten, a form of tungsten soft enough to be drawn out into filament wires without breaking. Tungsten proved better than carbonized filaments because it has the highest melting point of any conducting material suitable for drawing into filament wires.

Although Edison's large-scale searches might seem clumsy, recent episodes in the history of invention testify to the viability of the approach. In the late 1950s, Edward Rosinski and a colleague developed the zeolite catalyst for more efficient cracking of petroleum, systematically testing hundreds of combinations of conditions. New antibiotics are developed by assembly-line-like automation for culturing soil samples from various parts of the world. To the accumulating list of moves that figure in breakthrough thinking, it's worth adding the workhorse of the long systematic search, Edison's 99 percent perspiration.

How Stoneflies Came to Fly

Analogy, logical connections, asking the right question, systematic search—moves like these make up the human practice of breakthrough thinking. But how does mother nature do it?

Leonardo da Vinci and the Wright brothers are recent

history. Human beings are Johnny-come-latelies in the quest for flight, which mother nature has invented at least five times. Eons before the Wright brothers thought and labored in their garage in Kitty Hawk, North Carolina, small dinosaurs developed feathers, took to the air and became birds. Pterosaurs, flying lizards with weblike wings, preceded them by many millions of years. Flying insects had been on the scene long before that. During the Carboniferous period, about 300 million years ago, huge dragonflies with 30-inch wingspans cruised the primeval landscape, generating the energy to keep aloft from an atmosphere rich in 35 percent oxygen compared with today's 20 percent. Bats account for two more occasions of the discovery of flight. There are two quite different kinds of bats, each with its own evolutionary history.

If flight challenged the ingenuity of human beings, it did no less for nature. Evolutionists have long pondered how ground creatures could gradually become adapted for flight. Mutations cannot leap from no wings to fully functional wings. Typical sequences of evolution follow the paradigm of the fin or flipper. For instance, seals evolved from land animals, with flippers an adaptation of forepaws. A slightly longer and better-webbed forepaw afforded better swimming, favoring the survival of the proto-seal until, hundreds or thousands of generations later, seals attained their current form. Evolution is a story of the amplification of incremental advantages over time. Yet what makes sense for flippers makes little sense for wings. Suppose a mutation creates a small dinosaur with webs of skin under its forearms. This is still not nearly enough to fly, so where is the incremental advantage? What, in short, is a step-by-step path from a ground-dwelling creature to a flying one?

How various creatures took to the air is hard to read

from the sparse fossil record. However, a certain kind of pond insect tells an evolutionary tale through varieties that exist today. The winter stonefly is a small innocuous insect that makes its home on the surfaces of ponds, walking on the film created by the surface tension of the water. Biologist James H. Marden writes of the stonefly as a lesson in evolutionary problem solving. Some species of stoneflies fly easily and often, others a little, others not at all. But one nonflying species, *Allocapnia vivipara*, shows a provocative intermediate adaptation: It sails.

Imagine the scene: A cool wind blows across the surface of the pond. *A. vivipara* proceeds about its business only to find a frog about to make a meal of it. *A. vivipara* raises its tiny not-quite-wings and catches the wind. All of a sudden, *A. vivipara* glides across the surface of the pond beyond the reach of that darting tongue, propelled by its sails.

A. vivipara encapsulates in its anatomy and behavior one plausible account of getting off the ground (or in this case off the pond) by increments. Early ancestors of winter stoneflies had small winglike appendages that were not strong enough to support their weight while aloft. Evidence suggests that these appendages were too small and flimsy for flight. However, sometimes the membranes caught the wind and allowed the stonefly to escape from predators or find mates, an incremental survival advantage. Thus selection for stonefly sails began. Over many generations, the wings-to-be became larger and stronger. Eventually, some species of stonefly could lift away from the water altogether, another kind of survival advantage. The development of true flying wings began.

Mother nature doesn't think in the human sense of the word. The "thinking" of biological evolution is a kind of blind search process, one that we'll examine deeply toward the end of this book. But if mother nature did think, what kind of thinking would she be doing about *A. vivipara?* Interestingly, engineering has a name for this kind of thinking: *repurposing.* Appendages developed originally for sailing got repurposed for flying. Repurposing is one of mother nature's favorite ways of making breakthroughs—fins to legs, forelimbs to wings, scales to feathers. And repurposing is certainly a conspicuous mechanism in technological development, as Stone Age knives became scrapers, spears became arrows, wheels became waterwheels, fireworks became guns, oscilloscopes became television sets.

Mother Nature Breaks Through

Mother nature may repurpose, but do we see the full pattern of breakthrough thinking in nature—the long search, little apparent progress, the precipitating event, some nonmental equivalent of the cognitive snap, and transformation? Arguably, yes!

The very idea of a breakthrough implies there is a trap to escape. In nature, some paths of adaptation are straightforward and almost always get followed. Virtually every swimming sea creature has finlike or flipperlike appendages—an easy evolutionary problem with no breakthrough required. However, other paths of adaptation are subtle and tortuous. Many creatures might take to the air, but only a few have done so in some 600 million years of multicellular organisms.

If human breakthroughs take only a few moments, what about breakthroughs on the evolutionary scale? Traditionally, evolution was thought to be a gradual process. Necks got longer, beaks sharper, fur thicker over eons of time. Species gradually became other species, a kind of extended morphing that would be visible only in extreme time-lapse photography.

But one contemporary view of evolution writes the script differently. According to the *punctuated equilibrium* theory, evolution occurs in bursts that punctuate long periods of stasis. A new species emerges relatively quickly after not much change for millions of years. The initial period of stasis involves minor adaptations—mere fine tuning. Then a precipitating event occurs. The precipitating event might be a cataclysm that almost destroys some species and creates the opportunity for others to take over their ecological niches. Alternatively, the precipitating event might be the chance arrival of a species in a new environment—a seed washed up on a new shore, a rodent family struggling across a mountain pass—where it can adapt and thrive. Under such conditions, rapid evolution occurs, mere minutes on the scale of geological time. The punctuated equilibrium view preserves the idea that evolution is incremental: Small dinosaurs still do not sprout wings in a single generation. However, the theory holds that the process of incremental change can proceed quite rapidly.

Presumably, such a pattern would apply to the stonefly. Early in the stonefly's development, there was a slow process of adaptation, refining the stonefly's appendages for more efficient sailing. But at some point the process took off. The sails began to have significant lift, and selection pressure for true flight began to operate, relatively rapidly developing wings.

Breakthroughs in evolution suggest that insight in human beings is only one case in a larger class of processes. The dynamic pattern of long stasis and sudden shift marks breakthroughs both human and evolutionary. Although human breakthroughs reflect the shuffling of ideas and evolutionary breakthroughs the shuffling of genes, at a deep structural level they might have much in common.

The Mainspring Question

How creativity works has been one of humankind's fundamental questions for a long time. Two and a half millennia ago, Plato looked to divine inspiration for an explanation: "A poet is a light and winged thing, and holy, and never able to compose until he has become inspired . . . for not by art do they utter these, but by power divine." Especially puzzling are breakthroughs big and small. While more incremental versions of creative thinking have accomplished much of worth, it is the breakthroughs that draw our attention. And rightly so. Breakthrough thinking has made possible moon walks, penicillin, the Sonata form, Newton's laws, Impressionism, instant photography, Beethoven's late string quartets, and so much else. To understand breakthrough thinking would be to understand a significant part of what makes human beings successful as a species. It would also be to reach for more of that success, since—as I have argued in *Outsmarting IQ*— human intelligence can be cultivated.

So far, we have profiled what breakthrough thinking looks like on the surface: the long search, little progress, the precipitating event, the cognitive snap, and transformation. On the table are a few moves that contribute: analogy, logical extrapolation, asking the right question, systematic large-scale search, and repurposing. But these ideas do not penetrate very deeply nor advise very richly. We need an account of how breakthrough thinking works on the inside. One might compare the situation with examining a watch. It's easy enough to examine how a watch works on the surface—the hour hand, the minute hand, the numbers, and so on. But how the watch works on the inside is not so apparent. We have to pry it open to discover the mainspring, the gears, the escapement, and other gadgets that make the watch work. So what gadgets make breakthrough thinking work the way it does?

Psychologists studying problem solving and creativity, historians of science and the arts, paleontologists concerned with human evolution, and other sorts of scholars have pondered this question. One kind of answer proves more common than any other: Special mental processes do the work. These processes only swing into operation now and then, and perhaps only in very gifted people. For instance, some scholars write about incubation, a mental mechanism that supposedly solves problems for us while we pay attention to something else, like mowing the lawn. Others write about processes in the brain that sometimes make surprising connections or that alert us to crucial clues in medical or scientific mysteries.

These psychological accounts illuminate some aspects of breakthrough thinking, but there's a risk of giving them too much credit. The aim of this book is to put forth a very different view of how breakthroughs work. The idea in its simplest form says this: The surface pattern of breakthroughs reflects not underlying mental processes but the underlying structure of the problems themselves. To put it roughly, many problems are *reasonable*: They can be reasoned out step by step to home in on the solutions. But certain problems are *unreasonable*: They do not lend themselves to step-by-step thinking. One has to sneak up on them.

All this has an evolutionary interpretation too. Of course, evolution does not reason in the human sense, instead making a blind search through myriad rolls of the genetic dice. Even so, evolution arrives at some of its results in a fairly steady step-by-step way. However, as discussed earlier, other evolutionary results emerge by circuitous and relatively sudden routes, the equivalent of sneaking up on a solution. Flight is a good example of such a result, both for human inventors and for evolution. History shows that it was not a reasonable problem for humankind, and paleontology shows that it was not a reasonable problem for evolution. But both humankind and evolution were up to solving this unreasonable problem.

All this only previews ideas developed much more rigorously in later chapters. The remainder of Part I explores further examples of breakthrough thinking, drawing concepts from research on artificial intelligence to analyze the difference between reasonable and unreasonable problems. There is an art to approaching unreasonable problems, so Part II explores the strategic side of breakthrough thinking, drawing on a range of historical examples and puzzles. Part III looks at psychological research on insight, examining some of the mental mechanisms that have been proposed to explain breakthrough thinking, discussing their limits, but drawing real value from them. Finally, Part IV returns to the theme of evolution, teasing out the similarities and differences between how blind evolution and visionary human minds achieve breakthroughs.

2

From Sufi Tales to James Bond Thrillers

How to Be Blind

For centuries, Sufi mystics have instructed their disciples with stories. Many Sufi tales have passed into folklore, legend, and ethical teachings. The tale of the three blind men and the elephant is a celebrated example. Three blind men examine an elephant, seeking an understanding of this huge, ponderous object. One feels an ear and concludes that the elephant is huge and flat like a rug. Another feels the trunk and concludes that it is like a living pipe. A third feels its feet and legs and decides that it is like great pillars.

This story concerns a lack of breakthrough thinking. The tale underscores the difficulty of grasping the whole by sampling a single part. Stories, puzzles, and jokes that touch on breakthrough thinking are common in human history. Just as historical breakthroughs such as those of Darwin's and the Wright brothers can teach us much about breakthrough thinking, so can stories, puzzles, and jokes. They honor and celebrate the phenomenon, they can clarify the difference between breakthrough and ordinary thinking, and they may even teach us some skills.

The Islamic tradition of Sufi teaching tales entertains,

but also instructs. The tales aim to amplify the powers of perception and help disciples gain a deeper knowledge of God. In the Sufi tradition, students customarily immerse themselves in the stories chosen for study. The teaching master unlocks the internal dimensions of the tales when he feels a student is ready. One Sufi master said, "Only a few Sufi tales can be read by anyone at any time and still affect the 'inner consciousness' constructively. Almost all others depend upon where, when and how they are studied. Thus most people will find in them only what they expect to find: entertainment, puzzlement, allegory."

Here is another tale in the Sufi manner, and another story of breakthrough thinking.

One early morning, the traveler brought to the border between two nations a wheelbarrow full of straw. The border guard's suspicions ran high: There was no tax on straw, but what lay buried underneath? The border guard meticulously stirred through the stiff yellow stems, but nothing was to be found. Puzzled and exasperated, the guard waved the traveler on his way.

The next day the traveler appeared again with a wheelbarrow full of manure, another untaxed commodity. The border guard thought he saw through the traveler's stratagem. Stirring through straw was no problem, but manure made the guard's hands itch to be somewhere else. Still, the guard knew his duty. With a trowel, he investigated the noxious load, but again no contraband emerged.

Each day before the sun lifted above the buildings opposite the customhouse, the ritual was repeated. One time it was wood chips, another a load of gravel, another manure again. The searches grew into a friendly ritual.

"I know you must be smuggling something. I'll find out," the guard would say to the traveler with a grin.

"You've established my honesty many times over," the traveler would reply. The traveler was a cheerful fellow, and during the searches they would talk about the affairs of the day—who had swindled whom, the latest gossip about national leaders, and even the tricks of smugglers who now languished in the local prison.

"I'd hate to see that happen to you," said the border guard.

"An honest man has nothing to fear," responded the traveler.

All this continued for more than a year. Then one day the traveler failed to arrive with the sun and never appeared again. More than a decade later, when guard and traveler alike were living very different lives, they encountered one another in a tavern. The conversation focused on the here and now for a while. Then the guard asked the traveler the question that had been nagging him for such a long time. "I left that job years ago. I owe nothing to the government. And I know you were smuggling something. You must have been," he said. "For old time's sake, what was it?"

Puzzles as Psychlotrons

So what was the traveler smuggling? You may know the answer to the border guard's question because you have heard the story before. Or you may have figured out an answer, an insight that probably came with a sudden cognitive snap. Or you may still be baffled. In any case, the puzzle will return again in a new guise toward the end of this chapter.

But why puzzles at all, and why a chapter full of them? Because, as we inquire into the nature of breakthroughs and breakthrough thinking, it is difficult to re-create the experience of full-scale insights like Archimedes' discovery of the principle of displacement of water or Darwin's formulation of the theory of natural selection. One can report their history but not experiment with them. However, certain kinds of puzzles afford that opportunity. Usually called *insight problems*, they have an "either you get it or you don't" quality. They leave the problem solver flailing about with little progress until a moment of insight when a solution emerges. They create small-scale experiences of breakthrough thinking.

As such, they provide a tool of inquiry. Physicists probe the mysteries of atoms by bombarding them with particles using a cyclotron. Likewise, psychologists can probe the mysteries of breakthrough thinking by bombarding minds with certain kinds of puzzle problems. As physicists need a cyclotron, so psychologists need a "psychlotron."

However, there's a question of what's bombarding what. Recall that the core theory of breakthroughs in this book departs from typical psychological accounts. I'll argue that the characteristic pattern of breakthroughs—the long search, the sudden advance, and so on—reflects not unusual things certain minds do but unusual structures certain problems have. So maybe instead of bombarding minds with puzzles to see how the minds work, we are bombarding puzzles with minds to see how the puzzles work. You can have it either way you want. In any case, insight puzzles offer not only tools of inquiry but also a kind of mental gymnasium for cultivating the art of breakthrough thinking.

But not all puzzles call for breakthrough thinking. Some puzzles are what the last chapter called reasonable, inviting step-by-step approaches. Other puzzles are unreasonable, demanding breakthrough thinking. Breakthrough-like episodes of thought arise in additional contexts as well, for instance humor, which usually involves a sudden reorganization in our understanding of something. The following pages analyze the role of breakthroughs in reasonable puzzles, unreasonable puzzles, and humor.

Six Rompecabezas

Insight problem is a good descriptive term, but a more poetic one is *rompecabeza*. This lovely Spanish word comes in two parts. *Romper* is a verb meaning "to break," while *cabeza* means "head." Rompecabezas—head breakers. The name suits the occasion, because trying to solve such a puzzle is often like beating your head against a wall. The term offers a painful metaphor for two characteristics of breakthrough thinking mentioned earlier, the long search and little progress.

Here are six rompecabezas. The answers are given at the end of the chapter. Many more puzzles appear in this book, usually accompanied by strategies, hints, and explanations, but not this time. Simply try these out and see how you fare.

✐ The Coin
Someone brings an old coin to a museum director and offers it for sale. The coin is stamped "540 B.C.E." Instead of considering the purchase, the museum director calls the police. Why?

✐ Sahara
You are driving a jeep through the Sahara desert. You encounter someone lying face down in the sand,

dead. There are no tracks anywhere around. There has been no wind for days to destroy tracks. You look in a pack on the person's back. What do you find?

✎ The Mask

There's a man with a mask at home. There's a man coming home. What's going on here?

This is thin information, so perhaps you would like to hear some questions answered. Is the man with the mask a thief? No. Does the man coming home live there? No. Is the man with the mask going to hurt the man coming home? No.

Now, what's going on here?

✎ The Fan

This story was told to me as true. What's wrong with it?

Once many years ago during a long sermon, a man fell asleep and found himself dreaming of the Boxer Rebellion in China. In the dream he was captured and taken to the headsman's block. Meanwhile, his wife noticed he was nodding off. Just as the man dreamed the axe was descending, his wife reached over with her fan and tapped him briskly on the back of the neck to wake him up. The shock killed the man instantly.

Readers might have some questions along these lines. Does the answer depend on some obscure fact about the Boxer Rebellion that the story gets wrong? No. Indeed, the answer has nothing to do with the Boxer Rebellion. Could such a shock really kill a person? Let's suppose that it might. Did ladies take fans to church? Let's suppose so. There's a much more basic reason why, although this story was told to me as true, it can't be.

✏ The Rivals

Two sisters, daughters of a rich man, were ardent rivals in sports car racing, socializing, and the game of life in general. Their father got sick of their competitiveness and decided to teach them a lesson. He asked them to meet him at a deserted race track one day with their fancy sports cars. He announced, "The winner of the race will get a brand new sports car. But this is a race with a difference. The one whose car crosses the finish line *last* wins."

The two sisters hopped into the cars and roared off around the track as fast as they could go. Why?

✏ The Shot

One day a fellow carelessly handling a rifle accidentally shot himself in the head. Now, long as a rifle is, you might shoot yourself in the foot accidentally, but the head? Maybe a ricochet? No, the man was in the middle of an open field. How did he manage to do it?

Be Reasonable!

Puzzles like those above can prove annoying. Often people told the answers complain, "That's not fair! Be reasonable!" However, such problems are not so much unfair as different. They violate our expectations. We are used to problems that lend themselves to more sequential reasoning, where we can work our way to a full solution inference by inference. In contrast, insight problems are *un*reasonable in the most literal sense, not subject to sequential reasoning.

To get a feel for the difference, it's worth taking a

moment to "Be reasonable!" Here is an example of a reasonable problem from *100 Games of Logic* by the French puzzle maker Pierre Berloquin:

✐ A Three-Letter Word

Find a common English three-letter word, knowing that

- LEG has no common letter with it.
- ERG has one common letter, not at the correct place.
- SIR has one common letter, at the correct place.
- SIC has one common letter, not at the correct place.
- AIL has one common letter, not at the correct place.

(Again, answers to the problems appear at the end of the chapter.)

Such puzzles are genuinely challenging. However, they are not solved in a single cognitive snap. The answer emerges bit by bit through a chain of inferences. The first clue makes it possible to eliminate L, E, and G. The second clue reveals that one letter is to be found among E, R, and G—but from the first clue it cannot be E or G since they have been eliminated. Therefore, R is in the word, but not in the middle position. Continued reasoning in a similar style will quickly yield the solution.

Such a pattern of thought is very different from breakthrough thinking. The search moves forward step by step with steady progress. There is no precipitating event that triggers a cognitive snap. The moment of completion is hardly transformative, since it's just a matter of putting the last brick in place on the building. There may of course be mini-cognitive snaps along the way as the problem solver discovers the individual inferences that allow advancing, but no more of a breakthrough than that.

All this happens because we can steer by clear rules of inference that point the way to a solution. Cognitive scientist Margaret Boden, in her 1991 *The Creative Mind*, highlights the importance of such rules: "A merely novel idea is one which can be described and/or produced by the same set of generative rules as are other, familiar ideas. A genuinely original, or creative, idea is one which cannot." The need to change the rules from what they appear to be at first makes for unreasonable problems—head breakers.

There are many kinds of reasonable problems to satisfy the human relish for systematic reasoning. Psychologists Alan Newell and Herbert Simon foreground another in their classic 1972 book *Human Problem Solving*—cryptarithmetic problems, sometimes called cryptarithms. Here is an example:

✐ ABCs

The sum below is expressed only in letters. A stands for one of the digits 1 through 9, B for another, and C for yet another. Given the sum, what *must* A, B, and C be?

$$
\begin{array}{ccc}
  & A & A \\
  & B & B \\
\hline
C & B & C \\
\end{array}
$$

To reason out an answer, begin where a strong clue reveals one of the numbers. Here the letter on the extreme left provides such a clue: Where could the C come from?

Because there are no digits above it, the C has to be a carry from adding A and B. Since A and B cannot be larger than 8 and 9, their sum cannot be larger than 17, or 18 if there is a carry from the right-hand column. Therefore, C has to be a carry of 1 from the middle column: C = 1.

Now consider the middle column. Adding A to B yields B back again, plus a carry. To do that, A must be a sizable digit, perhaps 9. Yes, if A is 9 and there is a carry from the rightmost column, then 9 plus the carry is 10, which added to B would give B back again below. Therefore, A is 9. Now look at the right-hand column: A plus B has to yield C, which is already known to be 1. B has to be 2 to have that result. Therefore, A = 9, B = 2, and C = 1.

Notice how incremental this process is. To be sure, the clues may be hard to detect. There may be small cognitive snaps along the way as the clues emerge. However, the overall trajectory is one of step-by-step progress.

Here is another example of a cryptarithmetic problem.

✐ Three Names to Reckon With

As before, the letters stand for digits. Each letter stands for a different digit and of course stands for the same digit wherever it occurs. Given that D = 5, what digits match the rest of the letters?

$$
\begin{array}{cccccc}
D & O & N & A & L & D \\
G & E & R & A & L & D \\
\hline
R & O & B & E & R & T \\
\end{array}
$$

These and their kin can be fun and rewarding puzzles to work. However, they are still *reasonable* problems, problems that can be reasoned out. The realm of breakthrough thinking is the realm of unreasonable problems, rompecabezas where the usual methods fail.

The Hunting of the Snap

In *The Hunting of the Snark*, Lewis Carroll presented a mock epic about a search for the fearsome beast. The study of breakthrough thinking has its own Snark, the cognitive snap. Is the cognitive snap really there, and can we hunt it down and demonstrate its reality? The sample insight puzzles given earlier and the more conventional puzzles of the previous section make an intuitive case for the reality of cognitive snaps, which seem to occur much more with the first kind of problem than the second. Nonetheless, impressions might deceive.

Psychologist Janet Metcalfe of Dartmouth developed a simple paradigm in the mid-1980s to investigate this issue. As subjects worked on problems of various kinds, a tone would sound every 15 seconds. At that point, each subject would jot down a "warmth" rating on a 10-point scale to indicate how close a solution felt at the moment. A rating of 1 meant the subject felt very far from a solution, a rating of 10 meant a solution was in hand.

Employing this paradigm, Metcalfe tested a number of subjects solving insight problems. She found that episodes of problem solving where the ratings become progressively warmer—where subjects felt that they were converging on a solution—tended to yield mistaken solutions. In contrast, episodes of problem solving with a pattern of no increasing warmth and then a sudden leap to a 10 tended to culminate in correct solutions. In studies including insight and noninsight problems, Metcalfe and her colleague David Wiebe of the University of British Columbia found that progressive feelings of warmth predicted achieving a correct solution for noninsight but not for insight problems. In summary, the breakthrough-thinking pattern of limited progress with a cognitive snap at the end held true for insight problems and distinguished them from noninsight problems.

Janet Davidson of Lewis and Clark College conducted a study in similar style, contrasting subjects' response to problems classified in advance as insight and noninsight problems. Her findings replicated Metcalfe's results. Davidson also ran a variation in which subjects worked the problems with hints that helped them overcome the blocks the problems posed. The effect of the hints was to take away the insight character of the problems. Sometimes people solved them and sometimes they did not, but the pattern of warmth suddenly escalating at the end disappeared.

A rather different kind of finding also shows that insight problems have their own personality. For noninsight problems, how likely a problem solver is to solve a problem depends on how long the problem solver persists. Work longer, and you stand a better chance of reaching a solution. However, insight problems tend to be solved soon if at all. In a 1988 experiment, Robert Lockhart, Mary Lamon, and Mary Gick found that 84 percent of solutions to insight problems were produced within the first minute. Only 2 percent of the correct solutions were given after 2 minutes. It seemed that more effort only dug the same hole deeper, miring the problem solver further in the traps of the problems.

Insight on a Platter

Puzzles are not the only products of culture that often deliver a cognitive snap. Humor is another. Stephen Wright, a well-known stand-up comic from Boston, specializes in highly condensed quips, humor haiku, so to speak. One Stephen Wright joke goes as follows: "A man stopped me on the street and asked, 'Can you tell me the time?' I said, 'Yes, but not now.' "

Wright's joke and most jokes produce a miniature version of breakthrough thinking. The precipitating event of the punch line triggers a cognitive snap, producing a transformation of the situation. Consider Wright's punch line. A request for the time must be met now if at all. Yet, "Yes, but not now" reframes the request as though it were the kind of request that could be met later—a son or daughter asking for help with homework, an office mate asking for a 10-minute meeting.

The eminent cartoonist Gary Larson delivers the cognitive snap of humor with high reliability. A personal favorite shows a salesman peddling a program designed to teach thinking to a dim-looking fellow in the doorway. The guarantee of the program's efficacy, featured boldly on the cover of the product, reads "Double your IQ or no money back."

Like Stephen Wright's punch line, Gary Larson's mock guarantee suddenly restructures the little world of the cartoon in the mind of the reader. Those seven words provoke the integration of information in a few seconds. Consider all one needs to know to make sense of the cartoon: the idea that some guarantees are money-back, that IQ is a measure of intelligence, that a doubled IQ is an unlikely prospect, and that the audience for doubling your IQ is likely to be just those not-too-bright people least likely to see through a deceptive guarantee. All that can be spelled out laboriously in a paragraph, but the same complex integration of information occurs in a cognitive snap as people look at the cartoon.

Of course, not all humor functions like breakthrough thinking. Pie-in-the-face humor has more to do with disorganizing physiognomies than reorganizing worlds of perception. Nor is such conceptual reorganization all there is to humor. As Freud urged, humor often concerns cheating on social norms. Thus Wright's "Yes, but not now" refuses to comply with the social convention of polite response, while slyly offering the pseudocompliance of a deferred affirmative. Likewise, Gary Larson's salesman violates accepted practice with a cleverly misleading guarantee. Nonetheless, the commonalities between getting a joke and breakthrough thinking run deep.

However, there is a simple difference: humor is *packaged* breakthrough thinking. Like frozen peas, it doesn't taste quite like the real thing. Comedians craft their humor to stimulate a cognitive snap. Humor is a free cognitive lunch.

Even so, humor like the Sufi tales has enormous educative potential. Humor often reveals something about individual foibles and cultural oddities, helping people to understand themselves and others better. Marvin Minsky, a key figure in the development of artificial intelligence, suggests that humor advances cognitive acuity. Minsky argues in a provocative essay that, among other things, humor sharpens our awareness of conceptual boundaries. People only truly understand concepts when they can get and make jokes that deliberately play with the conceptual boundaries that set the concepts off from one another. Humor is a calculated boundary transgression that, like guerrilla forays, shows an understanding of where the boundaries are conventionally supposed to be by their very violation.

Half and Half

Thoroughly unreasonable insight problems, like the six puzzles presented earlier in the chapter, are a special breed. Naturally occurring problems that call for insight are usually mixed cases, partly unreasonable but partly reasonable. Some puzzles have this character too. A classic example of a mixed case, much used by psychologists and researchers on artificial intelligence, is this:

✐ The Missionaries and the Cannibals

Three missionaries and three cannibals are traveling together. They find their trek interrupted by a river. A boat cached in the bushes affords a way across, but the boat will only hold two people at a time. There is a danger: If the cannibals ever outnumber the missionaries on either side of the river, the missionaries will become dinner. What series of trips can transfer the group of six to the other side of the river without the cannibals ever outnumbering the missionaries on either side?

This problem calls for generating travel plans. Perhaps two cannibals take the boat across, then one comes back and takes a missionary across—but wait, that puts two cannibals and one missionary on the farther side. Another plan is needed.

Many travel options get eliminated at once because they lead to disaster for the missionaries. However, problem solvers often arrive at a cul-de-sac because of a tacit assumption. They presume that all the trips must take the form of two travelers over and one back. After all, how else can the travelers assemble on the other side of the river? However, this assumption is mistaken. There comes a point when two travelers who have already disembarked on the other side of the river on different trips must return together. Discovering this is a breakthrough, after which the rest of the puzzle yields easily.

This puzzle and others like it are mixed cases. Because of the insight in the middle, they score high on most of the insight criteria. On the other hand, solving the Missionaries and the Cannibals puzzle involves more than a sudden "Eureka!" So too for almost all genuine insights in the arts, the sciences, engineering, and even in the crime detection work of James Bond.

Goldfinger

The unfinished story from the beginning of the chapter has universal qualities of misdirection. One witness to this is a classic movie thriller that involves a trick very much like the one faced by the border guard. The James Bond film *Goldfinger* finds secret agent 007 working hard to expose the devious operations of Auric Goldfinger, international financier and dealer in gold in all its forms. On the surface a gentleman of good conduct, Goldfinger somehow smuggles gold out of England on a massive scale, reselling it in other countries at better rates. England's best border guards have no idea how Goldfinger manages to maintain this steady leak in England's pot of gold.

Goldfinger is nothing if not a high roller. He travels in the utmost luxury, in a beautiful old car chauffeured by his Korean bodyguard, Oddjob. He owns his own golf club in England, as Bond learns while edging out Goldfinger in a close game with trickery on both sides. Goldfinger warns Bond off with a deadly demonstration by Oddjob, who Frisbees his steel-rimmed bowler hat into a statue at the golf course, cleaving the statue's head off. Then Goldfinger leaves for the continent, luxury car and all elevated into a plane. Of course, gold might be concealed in the car. But no, officials have checked this before.

Using a tracking device planted on Goldfinger's car, 007 follows the financier to his secret smelter and, spying through windows, discovers how Goldfinger smuggles the gold. Parts of the car itself are gold. Goldfinger casts them in England, assembles them into the car, and disassembles and remelts them at his foreign smelter for further shipping. Not something in the car, but the car itself is the contraband.

The tale of the border guard and the traveler plays out the same scheme on a more rustic scale. The traveler's one-word answer to the border guard's question about what he had been smuggling was, of course, "wheelbarrows."

Solutions to the Puzzles

The Six Insight Puzzles

The Coin. Why did the museum representative call the police? If the coin were genuine, the makers of the *coin, working in 540 B.C.E., would not have known that Christ would be born.*

Sahara. *What was found in the pack on the man's back? An unopened parachute.*

The Mask. *What's going on with the man in the mask and the man coming home? A baseball game.*

The Fan. *What's wrong with the story about the man dreaming of the headsman's axe? If the man died instantly at the tap of a fan, no one would know what he had been dreaming about.*

The Rivals. *Why did both sisters race off as fast as they could? Each sister drove the other sister's car. Note the father's charge was "the one whose car crosses the finish line last wins."*

The Shot. *How did the man shoot himself in the head? The man discharged the rifle straight up into the air. The bullet nicked him on the way down.*

The Other Puzzles

A Three-Letter Word. *The answer CAR follows logically from the constraints.*

ABCs. *These letter-digit assignments follow logically:*

A	B	C
9	2	1

Three Names to Reckon With. *These letter-digit assignments follow logically:*

A	B	D	E	G	L	N	O	R	T
4	3	5	9	1	8	6	2	7	0

The Missionaries and the Cannibals. *Here is a travel plan that accomplishes the goal. The m's stand for missionaries, the c's for cannibals, and () for the boat.*

STEP	NEAR SIDE	FAR SIDE
1. Initial situation	m m m c c c ()	
2. Two cannibals cross.	m m m c	() c c
3. A cannibal comes back.	m m m c c ()	c
4. Two cannibals cross.	m m m	() c c c
5. A cannibal comes back.	m m m c ()	c c
6. Two missionaries cross.	m c	() m m c c
7. A cannibal and a missionary come back.	m m c c ()	m c
8. Two missionaries cross.	c c	() m m m c
9. A cannibal comes back.	c c c ()	m m m
10. Two cannibals cross.	c	() m m m c c
11. A cannibal comes back.	c c ()	m m m c
12. Two cannibals cross.		() m m m c c c

Notice how in step 7 a missionary and a cannibal who crossed to the far side on different earlier steps return to the near side together. If just one returned, the cannibals would outnumber the missionaries on the near side or the far side.

3

The Logic
of Lucking Out

Gutenberg Lucks Out

In the fifteenth century, there was no efficient technology
for printing. Books, including the Christian bible, had very
limited circulation. A metal worker in Germany, Johannes
Gutenberg, took this problem to heart and launched a systematic quest for ways and means to mass-produce the
Bible.

Gutenberg did not have to start from nothing. Printing
existed as a handcraft. Experts carved wooden plates and
used these to print many copies of a page. However, this
approach fell far short of what today would be called cost-
effectiveness. Carving a wooden plate was laborious and
time-consuming. Printing involved laying the paper against
an inked wooden plate and rubbing it enough to create a
good impression, also time-consuming. The carving-and-
rubbing approach constituted an advance over copies
hand-scribed by monks, but remained very much a cottage
industry with no capacity to deliver on a wide scale.

Gutenberg addressed two problems: putting together a
plate more quickly and printing it more efficiently. As to
the first, Gutenberg took a cue from stamps and seals of

the day that embossed their images on paper. He developed this idea into the technology of movable type, chunks of metal each carrying one letter that could be assembled into the text for a page. As to the problem of rapid printing, Gutenberg recognized the inherent inefficiency of rubbing the back of a piece of paper set against a plate. If only he could find a source of great pressure, then a single application of that pressure might take the place of the tedious rubbing, printing a page in seconds. But how could he generate such a force?

Taking a break from his fervent mission, Gutenberg attended a festival in celebration of the wine harvest. Quite by chance, Gutenberg encountered another technology that offered a clue. There amid the high spirits and abundant wine was the wine press used to squeeze the juice from the grapes. Gutenberg saw at once that the device contained the principles he needed to print a page in a single pass.

Gutenberg lucked out. He might not have gone to the wine festival, but he did. He might not have seen the wine press, but he did. This happens commonly with breakthrough thinking, when unexpected external circumstances provide the precipitating event that triggers a cognitive snap. It happened to Archimedes and Darwin as well.

But why did Gutenberg need to luck out? He was plainly a highly intelligent and committed individual, a person of perspicacity and persistence. Normally that would be enough, but it was not for Gutenberg, or Darwin or Archimedes, because the problems they faced resisted sequential reasoning.

Of course, more than luck was at work. The way Gutenberg had been working on the problem of printing the Bible greatly increased the likelihood that he would encounter a precipitating event. He was saturated with the problem. He had solved many aspects of it. Moreover, he had even asked the right question to finish the job, one of the key moves in breakthrough thinking identified earlier: How could one generate enough pressure to print a page in a single moment? The example of Gutenberg leads to two key questions:

1. Exactly what about the structure of insight problems resists sequential reasoning, leading to the need to luck out?

2. If the structure of a problem requires lucking out, what can be done to "up" luck—to systematically better the odds?

The Luck of the Klondike

An analogy helpful with these questions comes from an unexpected direction. In August of 1896, gold was discovered in the Klondike, a region of the Yukon Territory, Alaska. This quickly led to the gold rush of 1897–1899, a few fortunes, and considerable disappointment. Searching for gold in the Klondike bears a deep structural resemblance to puzzling over breakthrough problems with their unreasonable character.

The difficulty is that gold is where one finds it. The prospector knows what gold looks like, but cannot track gold down because gold does not leave tracks, except for occasional traces in riverbeds. The prospector has to spend a great deal of time casting about. Hunting for gold in the Klondike is challenging in at least four ways, and the same four apply to breakthrough thinking.

1. Wilderness of possibilities. There is a little gold in a lot of space. While the prospector has many places to look, only a few will reward the effort with the mother lode or even a handful of nuggets. The prospector has to struggle and persist simply to cope with the sheer magnitude of the task and the reality that only a few sites can be examined. This feature of the Klondike might be called a *wilderness of possibilities* standing between the prospector and the gold, or for short, a *wilderness trap*. Likewise, often in breakthrough problems there are many tempting directions but few actual solutions.

2. Clueless plateau. The second Klondike challenge is the lack of signs that point to the gold. Most of the time, the prospector samples gravel with at best only a trace presence of gold. Nor do the traces reliably point to more. Sometimes trace gold in the stream invites a harder look upstream, where the lode is, or downstream, where an accumulation may lie, but neither option may "pan out." This challenge of the Klondike also deserves a name: an apparently *clueless plateau*, or *plateau trap*. A plateau of low gold concentration with no obvious direction indicators to the deep pockets separates the prospector from the gold. Likewise, in typical breakthrough problems there are no apparent clues to point in the direction of a solution.

3. Narrow canyon of exploration. The third challenge is that the gold may be elsewhere altogether—over that mountain range in the next valley, at the headwaters of another stream, nowhere near the prospector's current efforts. Worse, the prospector, preoccupied with today's gravel pit in the middle of today's canyon, may not even be very aware of these other possibilities. This challenge might be called a *narrow canyon of exploration*, or *canyon trap* for short. Likewise, often breakthrough problems trap the problem solver with a taken-for-granted assumption or a limited representation of the problem or fixation on a habitual pattern of thinking. The problem solver searches vigorously for a solution, but within boundaries that do not contain a solution.

4. Oasis of false promise. The fourth challenge is that the prospector often finds the present locale too tempting to bother to look elsewhere. There are promising signs, perhaps even a little gold, enough to keep the prospector supplied and prospecting. The notion of walking away from this promising locale seems absurd. So the prospector continues to work the present site, in hopes that the mother lode lies just below the next shovelful of gravel.

The perfect word for this challenge comes from a setting very different from the frigid Klondike. This challenge might be called an *oasis of false promise*, or simply an *oasis trap*. In the desert, an oasis is not the destination but simply a stop along the way. However, it can be hard to leave an oasis and continue across the trackless sands with confidence of arriving at the ultimate goal. Likewise, often breakthrough problems tempt the problem solver with answers that are almost good enough, but not quite. It's hard to move away from them.

A Paper Klondike

The Klondike analogy cuts beneath the surface of things to disclose some fundamental structural features of breakthrough thinking. Any process of problem solving can be analyzed as a search through a space of possible approaches and partial solutions—a *possibility space* for the problem. The possibility spaces for different kinds of problems have different structures. In particular, breakthrough problems tend to have specific structural features that stand in the way of sequential reasoning—a huge wilderness of possibilities, clueless plateaus, narrow canyons of exploration, and oases of false promise.

Consider the following classic insight puzzle, an entire Klondike on paper:

✐ The Nine Dots

Draw four straight lines that pass through all nine dots in the accompanying diagram without lifting pencil from paper.

As usual, the reader may want to try this problem before continuing.

The Nine Dots problem has all the Klondike characteristics. There is a wilderness of possibilities, with many ways to try to draw the four lines, starting at different corners, proceeding in different directions. There is a clueless plateau: Few clues support a progressive search that homes in on a solution. One trial appears very much like another, and for the typical problem solver they all fail, with at least one dot dodging all four lines. Skipping to the fourth Klondike feature, oases of false promise, the near-solutions are tempting. It's easy to cover all but one dot, so perhaps adjusting a near-solution will capture all the dots. But it does not.

However, the peculiarly deceptive character of The Nine Dots problem also comes with the third feature, a narrow canyon of exploration. Most problem solvers spontaneously confine their lines to the box defined by the dots. The solution lies outside that area of search. The successful problem solver allows lines to extend beyond the edge of the box:

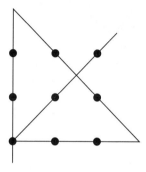

In general, problems that demand breakthrough thinking possess some combination of these Klondike characteristics—many places to look but few solutions (a wilderness of possibilities), few clues to point the way (a clueless plateau), the solution lying beyond a region of search one tends to fall into (a narrow canyon of exploration), and the temptation to persevere with near-solutions that won't quite work (oases of false promise). These four Klondike characteristics are what make such problems unreasonable—not amenable to sequential reasoning.

Of course, it's not that there are two utterly different kinds of problems, unreasonable and reasonable with nothing in between. The four Klondike traps can occur in various degrees. The difference between unreasonable and reasonable problems is somewhat like the difference between dangerous dogs and lapdogs: No sharp divide but a continuum separates them. Really dangerous dogs are near the big, tough, aggressive extreme, and lapdogs are near the small, soft, meek extreme, with everything in between.

The Four Operations of Breakthrough Thinking

The four Klondike characteristics reveal why lucking out plays a conspicuous role in breakthroughs. They also point directions toward thinking in ways that up luck, improving the odds. While unreasonable problems do not yield to sequential reasoning, they may yield when the problem solver accepts the Klondike character of the situation and proceeds with a kind of *Klondike logic*. Good breakthrough thinking systematically jumps the tracks of sequential reasoning.

Breakthrough thinking is an art and a craft, fueled by an inquiring spirit and abundant experience. Nonetheless, it can be systematized somewhat to bring out its basic patterns and strategies. Breakthrough thinking can be organized into four Klondike logic operations, one for each of the four challenges. They might be called *roving, detecting, reframing*, and *decentering*.

All four can be applied to The Nine Dots problem. Consider first *roving* far and wide. The Nine Dots problem presents a wilderness of possible ways to draw the four lines. *Roving* means exploring the possibilities widely, trying this and that. The spirit of far-ranging exploration might even lead one to try lines outside the box. Roving also often involves getting systematic about surveying all the possibilities. You can take advantage of the symmetries among the nine dots to compress the space of possibilities. There are really just three places to start: the center dot, a side dot, or a corner dot. Because of the symmetries, it does not matter logically which side dot or corner dot you choose. So there are really only three starting places. In fact, there are only two. The center dot is not an additional start, since the problem solver would naturally connect it to a corner or side dot as part of the first line, so the problem solver might as well have started with a corner or side dot in the first place.

Now consider *detecting* hidden clues in response to a seemingly clueless plateau. As you try different ways of drawing the lines in The Nine Dots problem, a problem of cluelessness arises. None of the approaches works, and it's not clear in what direction to search. *Detecting* means looking harder for clues that point a direction. For instance, you might carefully reexamine the problem statement for exactly what it says and perhaps discover that nothing was said about staying within the box. In breakthrough puzzles, it's often the absence of something that's a clue—what's not said that might have been said.

Now consider *reframing* the situation to cope with a narrow canyon of exploration. After a period of fruitless work on The Nine Dots problem, you are likely to notice that you are going in circles, trying the same things over and over. Then you might ask yourself, "What constraints am I taking for granted?" and abandon the frame within which you've been searching for a more liberating one. You might, for instance, observe that you've been drawing lines entirely within the box and adopt a more spacious frame.

Now consider *decentering* to avoid an oasis of false promise. With The Nine Dots problem, there are a number of ways of drawing four lines that cross all but a single dot—almost but not quite. It's possible to get stuck on these, trying to make them work with little variations. To decenter is to move away from seductive approaches that don't really work.

These four operations all illustrate the point that a breakthrough style of thinking only ups luck. Neither roving, nor detecting, nor reframing, nor decentering guarantees that someone addressing The Nine Dots problem will hit upon the idea of moving outside the boundaries of the box. But all of them improve the odds.

Here is a more formal account of the four operations of Klondike logic. Part II of this book will devote a chapter to each of them.

WILDERNESS OF POSSIBILITIES

Symptom: Overwhelming. There are a large number of possibilities, few of which are likely to pay off.

Response: Roving far and wide (for example, wandering all over the Klondike, or trying a wide range of possibilities in the Nine Dots problem). The idea of roving is to move around widely in the space of possibilities, looking here, looking there, not lingering long in any one place. Brainstorming, which will be discussed in a later chapter, is a typical kind of roving.

Roving can be casual, but it can also be systematic. The problem solver can search in ways that are thorough and avoid duplication of effort. It's also often possible to reduce the apparent size of the space to be searched. Sometimes the problem solver can capitalize on redundancies in the space to search only part and in effect get the whole, as in taking advantage of the symmetries in The Nine Dots problem. Sometimes the problem solver can eliminate a large group of unlikely possibilities at once, focusing on those that remain.

SEEMINGLY CLUELESS PLATEAU

Symptom: Cluelessness. There are no apparent clues pointing in a promising direction.

Response: Detecting hidden clues (such as looking not for trace gold but landforms that signal possible presence of gold in the Klondike, or looking for what rules of the game the statement of The Nine Dots problem does not address explicitly). Although there do not appear to be clues, perhaps there are, if only one looks in a different place or looks more carefully for incongruities or other suspicious features. Thus, in The Nine Dots problem, rather than just looking at your diagrams, it's worth looking back at the problem statement for clues. There the absence of specific constraints about staying inside the box can be seen as a clue.

NARROW CANYON OF EXPLORATION

Symptom: Confined. The problem solver is going in circles, trying the same range of things over and over. This suggests that the search may be unnecessarily circumscribed.

Response: Reframing the situation (for example, recognizing you are prospecting within one valley in the Klondike and trying out others, or asking yourself what assumptions you are making in The Nine Dots problem and finding you're operating within the square). When a search appears to be going in circles, the problem solver should investigate how tacit assumptions, descriptions of the situation, and other factors are constraining the search within a limited region. It's often effective to reframe the situation—to challenge key assumptions or represent the problem in a new way, broadening or shifting the boundaries of the search.

OASIS OF FALSE PROMISE
Symptom: Beguiled. A suspiciously easy solution or a partial solution or kind of solution that appears natural and inevitable beguiles the problem solver into persisting with it.

Response: Decentering from false promise (for example, leaving a low-yield claim behind in the Klondike, or abandoning Nine Dots near-solutions rather than struggling to make them work). This means recognizing the oasis for what it is and leaving it behind. A problem solver might do that by backing up to an earlier point and taking a different path from there, by simply bracketing off the current approach and trying something else, or by broadening or changing the problem definition to get away from an old solution. Sometimes the oasis can be a good point of departure, as long as one truly departs. The problem solver might discard some of the partial solution, keeping part and building on it, or reverse aspects of the current approach to try the opposite.

Notice how there is a certain sameness to the four operations. They all involve finding fresh possibilities. Moreover, each can help the others. For instance, by chance, roving far and wide may lead one to *detect* a previously hidden clue, *reframe* the problem in a new way, or simply *decenter* from false promise. But the four operations are not the same thing. They direct effort in different ways. A focused effort to detect, reframe, or decenter is more likely to accomplish these goals than simply roving. Likewise, reframing the situation may help with roving, detecting, or decentering but does not do their jobs as reliably. The four operations of breakthrough thinking function together like a team to address the Klondike challenges, whether the problem concerns printing the Bible, drawing four lines through nine dots, or changing careers as in the next example.

Everyday Breakthrough Thinking

The world of puzzles offers a kind of gymnasium for breakthrough thinking, but the true payoffs lie in the real world, where unreasonable problems arise in abundance. Imagine Andy, a thirtyish man suffering from chronic dissatisfactions with his current job. Andy wants something different, but what? As he begins to explore the possibility space of his problem, he realizes that his background prepares him for a variety of professions. Although this may not seem much like The Nine Dots problem or insight-demanding scientific challenges, at the level of ordinary life it is. Andy has entered a Klondike wilderness of possibilities. Andy can up his luck with systematic breakthrough thinking.

Andy roves the wilderness. Andy might get systematic. He could make lists, consult sources for ideas, group careers of similar kinds, cover the space of possibilities.

Andy might narrow the field drastically without any real loss by excluding whole categories of professions for which he truly lacked a flair, even if they appeared superficially attractive. Of course, Andy would not want to pick criteria so narrow that they simply lead him back to his present job.

Andy detects his way off a plateau. Suppose the new jobs Andy considers all seem okay but not ideal. They don't point a clear direction toward better possibilities. Thus, Andy finds himself on a seemingly clueless plateau. Perhaps he needs to look more deeply to detect the clues. He examines his work experience carefully. "When am I feeling good? When am I feeling not so good?" He might find that it's really working relationships that make the difference. Okay, so he needs to build better relationships in his current setting or maybe start over in a new one more supportive of good relationships.

Andy reframes his way out of a canyon. Andy might find that he's going in circles. "Job A looks good, but there's that impossible commute. Job B would be different and fresh, but I don't have the background for it. Job C—attractive, but the pay is poor. My present job isn't really so bad. But what about A again? Well, let's see . . ." So to break out of the canyon he asks, "What assumptions am I taking for granted? Maybe I'm assuming that my work is what's bothering me." Lifting his vision above work, he discovers that what troubles his life is not really his work at all but a chronic health problem or worry about aging parents. His focus on work problems is a trick of his mind to avoid facing the real problem.

Andy decenters from false promise. Andy finds his thoughts fixated on an attractive but tricky new professional direction. He keeps trying to figure out how to make it work: "How could I find enough to invest? How could I find a partner? How could I get the background I need?" To decenter from this tempting but intractable solution, Andy brackets it off for a while and explores altogether different possibilities. Maybe he'll come back to it, but not now.

Six Paper Klondikes

The following six insight puzzles all ask for systematic breakthrough thinking. A problem solver's thinking will tend to be limited by tacit assumptions that need to be questioned (narrow canyon of exploration) and by tempting near-solutions that need to be bracketed off to progress (oasis of false promise). There may be a clueless plateau or two as well. Readers are invited to try these puzzles with breakthrough thinking in mind to up their luck. To help the process along, the section after the problems offers hints about how to apply breakthrough thinking, with the full answers at the end of the chapter.

Don't be discouraged if these problems prove tricky despite the four operations. It is not easy to move from principles to practice. The four are not a magic wand. They have to be applied with care and persistence. More than anything else, getting familiar over time with the style of such problems helps.

✏ The Two Strings

This is a version of a classic insight problem used extensively in psychological research. You face two strings hanging from a ceiling some distance apart.

Holding one string in one hand, you still can't quite reach the other string with your other hand. Your task is to tie the two strings together.

To work with, you have the following objects: a dictionary, a stapler, a glass, a live toad, and a clothespin. How might you tie the strings together?

✎ To Tell the Truth

Two strangers meet at a party and fall into a conversation about their lives. At one point, the first stranger says, "I have to confess that I don't always speak the truth."

The second stranger replies, "Well, that I certainly must believe." Yet the second stranger has not heard the first say anything he knows to be false. Why is the second stranger so sure the first stranger's confession is sound?

✎ The Joke

One day at the office, Alice says to Betty, "I heard this great joke from Cathy." And she begins to tell Betty the joke.

But Betty says, "Oh, I already know that joke."

Alice says, "Oh, Cathy already told you."

"No," says Betty. "In fact, I never heard it or read it before."

Explain how this could be.

✎ The Extended Family

Four relatives spend a very pleasant day together catching up on old times. Although there are just four of them, all by themselves they make up an extended family. The four include a father and mother, a son and daughter, a sister and brother, an aunt and uncle, a niece and nephew, and two cousins. All of these relationships occur within the group of four (the father is a father of someone else in the group, and so on.) No marriages of people already relatives are involved. How is this possible?

✐ The Fly

Two people stand 10 feet apart. They begin to walk toward one another at a smooth steady pace of 1 foot every 10 seconds.

A fly sits on the nose of the first person. As they start to walk, the fly takes wing and flies to the nose of the second, then back to the nose of the first, then back to the nose of the second, and so on. The fly's speed is 1 foot per second.

Of course, the distance for each trip is shorter, because the people are closer together. How much distance does the fly travel before being crushed between the noses of the two people as they run into one another?

✐ The Equation

Here is an equation:

$$2+7-118=129$$

As it stands, it is not a valid mathematical statement. Your challenge: Add one straight line anywhere in the equation to make it a true statement.

This puzzle has at least three very different solutions. Try to find all three!

Six Hints

How did you do? Only so-so? Well, systematic breakthrough thinking is a trap shrinker, not a trap eraser. What is likely to help is examples and practice. The following comments avoid revealing the answers to the six insight problems but offer some hints by making connections with some of the breakthrough thinking principles.

The Two Strings. Psychologists have used The Two Strings problem to investigate *functional fixedness*, the tendency to see objects in their normal functions but not in less usual functions. When people use a dime to turn a screw or a lipstick to leave a message on a mirror, they are breaking away from functional fixedness.

Functional fixedness creates a narrow canyon of exploration: Problem solvers only consider typical functions of the objects available. This suggests recognizing the boundary and reframing more broadly. In the case of The Two Strings puzzle, it's worth ignoring the normal functions of the objects provided. How might one of them be used very differently? Which object? Not the toad, which is a distraction, a kind of small oasis that tempts one to attend to it because it is unusual and seems to be a clue.

To Tell the Truth. When the first stranger admits, "I don't always speak the truth," it is natural to wonder about the statements he might have made earlier in the conversation. Unfortunately, none of them are revealed and the second stranger states that he does not know any of them to be false. Therefore, there are no clues in the form of statements the first stranger makes earlier. Problem solvers find themselves on a plateau of seeming cluelessness with nowhere to go. But let's do some sharper detecting of clues: Consider as a clue the one explicit assertion the first stranger did make. Could *that* statement be true?

The Joke. Almost always, people hear jokes and pass them along to others. Jokes are like money, a medium of exchange that flows from person to person without being created or destroyed. However, thinking of jokes only in this way is a narrow canyon of exploration. Let's broaden the frame: Jokes have to come from somewhere in the first place.

The Extended Family. This problem has a tempting oasis of false promise in it. As the first several relationships are read—father and mother, son and daughter, sister and brother—it's natural to think, "Aha, a couple with a son and a daughter, who are of course thereby sister and brother." Unfortunately, this does not leave room for the aunt, the uncle, and two cousins. Decenter from this oasis and consider other possibilities.

The Fly. This problem too includes an oasis. The temptation is to think about the chain of trips of decreasing length the fly makes. If only one could sum up all those trips! This problem is especially entrapping for people with some mathematical background, because there are various strategies for finding the sum of sequences of systematically changing numbers. Instead, it is best to decenter, avoiding the idea of adding up all the trips. Perhaps there is another way altogether to determine how far the fly travels.

There is a narrow canyon of exploration here also. The problem asks about the distance of the fly's trip, so it's natural to think in terms of distance. Perhaps one can reframe and think in terms of a different parameter than distance. For instance, total flying time.

The Equation. This puzzle has an oasis of false promise

and two narrow canyons of exploration. The oasis comes from the simple but tempting change of the minus to a plus. After all, it is the minus 118 that makes the equation so far from being true. With a plus, the left side sums to 127, just 2 short of the 129 on the right. If only another change were allowed—to add in the 2 somehow—but the one allowed change has already been used. So decenter from that way of thinking, and try other approaches altogether.

The first and easiest narrow canyon of exploration to escape from is the idea that the expression after adding a straight line must be an equality. However, did the problem statement say so? No. So broaden the frame. The only constraint is to make the expression a true mathematical statement.

However, there is a solution with an equality, protected by another canyon. The solution proves difficult to discover because of a natural canyon tendency to explore what lines might be added, either to (a) change numbers into other numbers, or (b) change operations (+, -, . . .) into other operations, or (c) change relations (=) into other relations. Might it be possible to reframe and change an item from one of these categories into an item from another one of them?

Six Solutions

Here you can check your solutions. If even with the hints you had a hard time, don't worry about it. Seeing how these examples work will help you with later ones.

The Two Strings. *The classic solution to this puzzle is to tie a medium-weight object—say the stapler (the toad is included for the sadists)—to one string and set it swinging. Then, holding on to the other string, the person grabs the first string on its upswing and ties them together. As* mentioned earlier, functional fixedness is a canyon trap that blocks this solution. One tends to think of the objects in terms of their normal function. By broadening this frame, one can think about other functions.

To Tell the Truth. The apparent lack of clues creates the plateau, because of not knowing what the first stranger said before his final statement. However, the one assertion he did make offers a subtle clue one might detect: "I don't always speak the truth." This statement has to be true. Suppose it was false. This implies that the stranger does always speak the truth. However, he says he does not, which contradicts the idea that he does. Thus the supposition that his statement is false leads to a contradiction, demonstrating its truth.

The Joke. The canyon here is the notion that jokes just get passed along from person to person. But let's broaden the frame. Someone has to make them up in the first place. In particular, Betty composed the joke and told it to Cathy, who in turn told it to Alice. So, when Alice began to tell it to Betty, she already knew it without ever having been told.

The Extended Family. The oasis is the temptation to make the father and mother, the first-mentioned family members, a married couple with their son and daughter who are sister and brother. However, this combination does not meet the conditions of the rest of the puzzle. Decentering from that temptation, here is a different configuration: a sister and brother, each of which is married to someone not present, with the sister's daughter and the brother's son (or sister's son and brother's daughter). In that case, the sister is sister to *the brother and mother to the daughter. The daughter is cousin to the son and niece to the brother, who is of course uncle to her, and so on.*

The Fly. *The oasis here is the temptation to try to add up somehow the successively decreasing distances of the nose-to-nose flights of the fly. A canyon is to think just in terms of distances. Decenter from adding up the decreasing distances, and reframe to consider not distance but total travel time instead. The people, starting 10 feet apart, approach each other at a rate of one foot every 10 seconds. This means they run into one another when they have each traveled 5 feet, in 50 seconds. This in turn means that the fly has been traveling for 50 seconds at a rate of 1 foot per second. The fly's total distance before its final nose dive: 50 feet.*

The Equation. *The first two solutions come from recognizing that the expression need not be an equality, the first canyon trap. One can simply put a slash through the equals sign to make it "does not equal," or one can put a diagonal line upward from the right end of the equal sign to make the expression read "less than or equal to." Both of these changes create true mathematical expressions as required by the problem.*

The third solution involves reframing more broadly to escape the canyon trap of changing operations to operations, relations to relations, and numbers to numbers. The + can be changed to a 4 by adding a vertical line on the upper left of the sign. This makes the equation true.

These examples make for a good moment to address a common question. Canyon and oasis traps seem somewhat alike. How in the end do they differ?

Broadly speaking, it's a matter of getting out versus getting away. We want to get out of the canyon and away from the oasis. Getting out of the canyon is a matter of identifying the boundaries that limit the scope of search and then reframing them—for instance, using objects in ways that go beyond their usual functions as in The Two Strings problem, or asking not just how jokes get passed around but where they come from in the first place as in The Joke. Getting away from an oasis is a matter of abandoning a tempting partial solution or near-solution or seeming solution that does not seem to be living up to its promise—for instance, discarding the partial solution of a married couple with their son and daughter in The Extended Family, or giving up trying to fix the sum that almost works in The Equation.

This said, there are certainly borderline cases that one could describe either way. That's not surprising. Borderline cases occur all the time with familiar categories, for instance, a chair with a low back and a high seat—is it a chair or a stool? But such borderline cases do not leave us feeling that there's no distinction between a chair and a stool, and neither should borderline cases between canyon and oasis traps leave us feeling there's no distinction there. From a practical standpoint, when borderline cases do occur, either reframing or decentering might work just as well and yield much the same results.

4

Is There a Science of Breakthrough Thinking?

Alchemy or Chemistry

The medieval alchemists imagined that one substance could be made into another with the right meld of ingredients and incantations. Their Holy Grail was the conversion of lead into gold, perhaps with the help of a touch from the transformative philosopher's stone. Their experimentation laid the foundation for the science of chemistry, but, in their era, what later became chemistry was an odd mix of messing around, magic, and hope.

Today, when we analyze breakthrough thinking with an odd mix of concepts like unreasonable problems, possibility spaces, and Klondike challenges, is this alchemy or chemistry? The case can be made for chemistry. The scientific side of the story comes particularly from a body of research on artificial intelligence developed over the past fifty years. Artificial intelligence concerns efforts to model and produce intelligent behaviors such as problem solving, game playing, and planning by programming computers. While there are different approaches to artificial intelligence, one of the most prominent foregrounds the idea of searching through a space of possibilities. This approach applies not only to computers but to human beings as well. Computers programmed in this way do not think as well as people, but what artificial intelligence programmers need to do to get them to think even a little like people illuminates how human thinking works.

This research tradition offers a set of concepts and language that can make ideas about breakthrough thinking rigorous. The concept of possibility spaces provides a formal way of defining reasonable versus unreasonable problems, and it offers a framework for contrasting what kind of thinking is smart in the one and the other case. The rigor brought to breakthrough thinking will not do much to make the practical art clearer. Readers who take that as their principal concern are welcome to skim or skip. But the foundation is at least reassuring. We are closer to chemistry than alchemy when we look at breakthrough thinking in terms of its Klondike characteristics. The gold is found not by the magic touch of a philosopher's stone but by definable strategies of smart search.

Everything Is Like a Game of Chess

Zelda and Uri are playing chess. Zelda has advanced a piece, and Yuri needs to select a response. Yuri realizes that he could place his knight here. Then, if Zelda did nothing threatening, he could advance his rook there. On the other hand, Yuri thinks, Zelda could attack the knight with that bishop. So Yuri had better look beyond the knight move. Perhaps castling would be a better plan. Eventually, Yuri selects a move that he hopes will lead to an advantageous position.

Yuri has solved a problem by searching through a space

of possibilities. Choosing a move in chess may or may not be a breakthrough problem, depending on the position, but it is certainly a problem. For any kind of problem, whether competitive or solitary, whether a game or serious business, the notion of search through a space of possibilities provides a powerful way of conceptualizing problem solving. People look at possibilities, explore what other possibilities they lead to, and even what possibilities *those* lead to, eventually making a choice they think will help get them where they want to go. In breakthrough thinking, people break through to new areas of that space of possibilities. Everything in the world of problem solving is like a game of chess—what moves do I have, what moves might lead to what, and so what should I do?

The idea of searching through a possibility space, often called a problem space, lies at the heart of the classic *Human Problem Solving*, a 1972 book by cognitive scientists Alan Newell and Herbert Simon. The authors analyzed several different kinds of problem solving using this concept of problem or possibility spaces. Choosing a move in chess was one. Another was the cryptarithmetic puzzles introduced in Chapter 2 that presented addition problems with the digits represented by letters, such as DONALD + GERALD = ROBERT. Newell and Simon collected think-aloud protocols of people working on such problems and compared them with computer simulations of search in appropriate possibility spaces. They gathered evidence that people indeed thought in this exploratory way, navigating through various possibilities in search of ones that advanced toward a solution.

Search through a space of possibilities is a concept that can be made quite rigorous for many games, puzzles, and mathematical problems. The key characteristics of such a formal model will be useful for the arguments in this chapter. Here they are, with chess and cryptarithmetic problems as examples:

The state space. Also called the possibility space, the state space is the set of possible states of affairs, "states" for short. The state space for chess consists of all the legal configurations of pieces on the board, whether strong positions for the player or not. The state space for a cryptarithmetic problem consists of all possible ways of associating digits with some letters (not necessarily all the letters, because one may only be part way through the problem).

Operators. These are actions that shift from one state to another. The operators for chess consist of the legal moves, each of which transforms one chess position into another. The operators for cryptarithmetic problems consist of adding a digit-letter assignment, or retracting one, or modifying one.

Initial states. There is an initial state. In chess, it is the standard starting position. In the cryptarithmetic problem, it is the state of no digit-letter assignments, before any problem solving begins. There can be more than one initial state. When cards are dealt in bridge or poker, this creates a fresh initial state for each game.

Solution states. There is a criterion determining solution states. In chess, the criterion is checkmate. A move chosen in the middle of the game, although it solves the problem of what to do now, is just a temporary solution to be evaluated by its value in bringing the player closer to checkmating the opponent. In the case of cryptarithmetic, a solution state would be any assignment of all the letters to distinct digits that makes the sum true.

Measure of promise. Finally, there is a measure, or indicator, of promise, of how close the state-so-far is to a solution state or whether the search is moving toward or away from a solution state. For chess, the indicator of promise would be the overall advantage of a board position for one player versus another. For cryptarithmetic, a simple measure of promise would be how many letters are assigned to digits without immediately violating the rules of arithmetic and the given letter sum.

The measure of promise creates what is sometimes called a fitness landscape. The higher the measure at a particular state, the more adequate the state. A search process can explore the possibility space, monitoring the measure of promise as a guide, climbing toward higher promise and ultimately a solution.

Of course, when people play chess or attempt cryptarithmetic problems, they are not thinking about states, operations, and measures of promise as such. They are playing the game, not theorizing about how they are playing the game. However, Newell and Simon demonstrated with their experiments that this is a good way to model what problem solvers are in effect doing. Whatever the subjective experience, problem solving amounts to a search through possibility spaces.

Taking Off the Tuxedo

Playing chess has strict rules, but designing a bridge does not. In cryptarithmetic problems the sums have to be right, but who's to say what's right when a poet writes a poem? The origins of the possibility space concept are formal, mathematical, rigorous. However, to adapt the concept to explaining problem solving in general, we need to stretch it to accommodate the messiness of informal situations.

Informal situations are messy in at least three ways. The space of possibilities to be searched may be fuzzy, with one state blurring into another rather than being neatly set off as in chess positions. The space of possibilities may appear to change during the course of problem solving, for example, as one gets new information. Finally, one's criteria for success may evolve as one thinks about the problem, rather than staying fixed from the beginning. In such circumstances, can the idea of search through a possibility space still provide a good account of problem solving?

Fuzzy Possibility Spaces

It's fairly straightforward to analyze problems with fuzzy possibility spaces in the language of states, operators, and criteria of promise and success. Consider a problem like rescuing a cat from under the crawl space beneath a cottage. What could the frustrated owner do? He might call the cat out. He might tempt it out with a saucer of cream. He might lure it out with a squiggly strand of yarn. He might let it come out in its own good time. As such options occur to the cat owner, he is exploring a space of possibilities and evaluating them for their promise. When the cat owner actually tries out one idea and then another, he is exploring a space of concrete physical possibilities— like actually moving a chess piece after thinking about what move might serve.

For another example, suppose an orchard owner wants

to design an apple picker, a device for reaching into a tree to snag apples. Perhaps the apple picker can have a hook on the end, to pull the apples off the tree. However, then the apples would fall to the ground and bruise. Perhaps the orchard owner can improve the design by slinging a bag at the end of the pole under the hook, to catch the apples as they fall. Interesting, but as the bag fills up the pole would become hard to hold up. Perhaps she can put a counter-weight at her end of the pole, to help keep the pole balanced as the bag fills. Perhaps she can sling a cloth tunnel from the top of the pole to the bottom, an apple-duct that guides the apples down into a bag at her side. As the orchard owner ponders these ideas, this too becomes an exploration of possibilities and selection according to their promise.

In addition to possibilities and promise, the ideas of states and operators apply. Imagine the cat owner trying to get the cat out from under the crawl space. The states are snapshots of the possible scenarios—the saucer of milk set out and the cat still in the crawl space, the cat licking up the milk, the cat in his arms. The operations take the cat owner from one state to the next—setting out the saucer of milk, the cat coming to the milk (the cat's operation, like the move of an opponent in chess), the owner scooping up the cat. For the orchard owner, the states are the different design ideas for an apple picker with their various features. The operations are whatever thoughts take the orchard owner from one design to another—adding features, subtracting features, modifying features.

Even activities like poetry writing or painting can be described in these terms. Poetry writing involves a state space of partial and complete poems—anything from a few words jotted down to fragments of stanzas to an entire poem. The operations are actions of adding and deleting words. A poet navigates through this space, adding and deleting, seeking a good poem. The measure of promise is not a formal rule but the poet's own poetic judgment, the solution state a complete poem felt to be satisfactory. Likewise, the state space for a painting is the space of partial paintings and paintings, anything from a few brush strokes to a complete work. The operations are ones of adding or scraping away paint or overpainting. As the painter searches for a compelling painting, the measure of promise and success is not a formal rule but the painter's own judgment as an artist.

At one level, the possibility spaces for poems and paintings are part of concrete reality, actual sketches on paper or canvas. However, the poet or artist also searches mental possibility spaces of ideas and images before wielding pencil or brush, just as the chess player searches mental possibility spaces of moves and countermoves before touching a piece. These mental possibility spaces, together with the actual moves on paper or canvas, make up the entire possibility space.

Changeable Possibility Spaces

Suppose the cat discovers mice under the crawl space. Now a dish of milk holds little allure. The mice have radically changed the possibilities for retrieving the cat. Such upsets are commonplace in problem solving. In chess or cards, an opponent makes a play that changes the situation radically. In the world of scientific inquiry, someone else publishes a related result that changes the shape of the problem one is working on. Possibility spaces in general are less like the relatively fixed topography of the Rocky

Mountains and more like the constantly shifting topography of Cape Cod dunes. It's worth recalling the familiar saying, "The ground keeps shifting under our feet."

All this might appear to threaten the very idea of possibility spaces. What sense does it make to speak of searching a space that changes all the time? However, there is another way to think of such shifts. They are not changes in the space but changes in the current state, initiated by outside factors. When Zelda makes a move in chess, this shifts Yuri's position within the space of possibilities he must deal with. When mice appear under the crawlspace, this carries the cat owner into a different region of the large possibility space of rescuing cats from crawl spaces, where the owner must seek different strategies.

In general, what may seem like a change in the possibility space can be thought of instead as a jump from one place to another within the possibility space caused by some factor not under one's control. Where one ends up has always been a possibility in the possibility space—it's just become actual. Rather than threatening the idea of possibility spaces, such jumps recommend exploring the possibility space more deeply, anticipating what unexpected jumps might happen. Thus, Yuri does well to foresee what radical moves Zelda might make, and the cat owner does well to foresee what surprise cat behaviors or mouse appearances might complicate his situation.

Evolving Criteria

Another kind of change that plagues the possibility space model concerns the solution criteria themselves. What if the cat owner decides that he would rather have the cat hunting down the mice under the crawl space than lounging in the living room? When people address *formal problems*, criteria for a solution are well defined from the beginning and stay constant—checkmate, or the right sum for the cryptarithmetic problem. However, in *open-ended problems*, criteria often evolve during the course of problem solving. Sometimes at the beginning there is only a vague conception of what the problem is—"I wish apple picking were easier!" or "Where's that cat?" or "What a great opening line for a poem!" Exploring different visions and versions of the problem makes up an important part of the thinking. Along the way, the problem solver becomes clearer about what the problem is and just what would constitute a solution. Perhaps the criteria even shift dramatically at some point. The orchard owner wants not a tool but a creature: trained monkeys picking apples. The cat owner wants the crawl space clear of mice. The poet wants not a sonnet but a sestina.

Although criteria that change along the way seem threatening to the idea of search in a possibility space, they are easily accommodated in the formalism. Instead of envisioning search through a space of possibilities with fixed criteria, it is only necessary to envision search through an expanded space of possibilities consisting of both the original possibilities *and* possible criteria. The problem solver explores the original possibilities and possible criteria at the same time, with a solution consisting in a good match. The search proceeds on two fronts at once—the current cat rescue plan or partial poem or provisional idea for an apple picker, and the evolving criteria for satisfaction.

In summary, search through a space of possibilities makes an appearance even if the space of possibilities is fuzzy, jumps occur from one part of the space to the other, or criteria evolve in the process. To be sure, the states, operations, and criteria for progress and success are not usually as neatly or overtly defined as in chess. But after all, chess and many other games are deliberately constructed in lean, logical ways to remove the fuzziness characteristic of real problem-solving situations. Whether the world at hand is the world of chess, cats, apple pickers, poems, paintings, scientific theories, bridge architecture, or recipes for bouillabaisse, the language of spaces of possible states, operations, and gauges of progress and success provides a conceptual system for examining the process of thought and contrasting the demands of different kinds of problems.

When Smart Is Reasonable

With all this as background, we turn to the contrast between reasonable and unreasonable problems. Both are a matter of search. In the one case, reasoning offers a smart way of searching, but in the other it does not.

The simplest search strategy is to examine all the possibilities. In some situations this is the right thing to do. Most people at some point master the familiar game of tic-tac-toe, either learning the ropes from someone older or figuring it out. I remember getting annoyed when my father kept beating me, so I sat down with paper and pencil and figured it out. When one takes into account the symmetries, there are really only three places to start—corner, side, or middle. For the next play, there are only a few other genuinely different possibilities. From then on, I could hold my own.

Examining all the possibilities is not always a trivial strategy. It requires attentive and systematic action to be sure not to miss possibilities. Nonetheless, for most problems examining all the possibilities is not smart strategy because there are too many to investigate. The DONALD + GERALD = ROBERT cryptarithmetic puzzle has ten different letters, standing for the ten digits 0 through 9. It's stated that D = 5, so there are nine remaining letters to assign to the nine remaining digits. Mathematically, this might be done in 362,880 different ways. A problem solver certainly would not want to consider all those possibilities. In general, in problems that are at all challenging, there are too many states and too few solution states to find solutions by searching thoroughly. What is needed is another style of smart search. Cognitive scientists speak of heuristic search, the term *heuristic* referring to strategies that increase the chances of success without guaranteeing success.

A basic strategy of smart search is to follow the measure of promise, tracking its increases through the fitness landscape to a solution. In the case of a cryptarithmetic problem, this means proceeding incrementally, starting with one letter-digit assignment consistent with the information given, then adding another, then adding another until all are in place. The key to this strategy is to use the logic of the situation. A smart search would try to find a digit-letter assignment forced by the given information—that is, one that could not be any other way—and then another and another, drawing out any immediate implications and exploring alternative branches only when necessary. When there are many possibilities, smart search examines far less than the entire space.

The strategy of following promise applies not only to formal problems but to fuzzy ones too. Consider the woman searching through different possibilities for an apple picker and how one step led to the next. The search was progressive: first a hook, but the apples would bruise on the ground; then a bag, but it would get unbalanced; then a counterweight, but that would be heavy too; then a cloth tunnel. In a loose sense, this progression is like solving a cryptarithmetic problem. The first idea contributed part of the solution, leading on to other adjustments and extensions that provided a more complete solution.

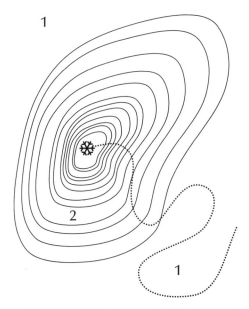

Search in a Homing Space: 1. Clueless regions.
2. Large clued regions leading to the target.

The accompanying figure illustrates in a general way the fitness landscape of a reasonable possibility space—one that lends itself to following promise. As the picture shows, such a possibility space has a relatively simple structure. The contour lines show regions where there is a discernible slope to the measure of promise. In those regions, the problem solver can follow the slope of increasing promise to the solution, homing in on it systematically. Such possibility spaces might be called homing spaces because of this convenient structure.

Is homing easy? Not necessarily. Although following promise is a reasonable approach, it can be a very challenging one. Following promise requires systematic attention and meticulous logic. Following promise means making the most of the given information. Solving a difficult problem of this sort is like climbing a cliff: The climber needs to capitalize on each little niche for a handhold or toehold.

When Smart Is Unreasonable

Challenging though the world of homing spaces can be, it is not the world of Leonardo da Vinci's or the Wright brothers' insights about flight, nor the world of Gutenberg's invention of the printing press, nor Darwin's discovery of natural selection, nor the Sufi tales and their hidden meanings, nor The Nine Dots problem. These problems and puzzles have unreasonable possibility spaces—Klondike spaces. A visualization of a Klondike space appears in the next figure.

Think of the space as huge with, in this case, only one solution in the top right. The size and rarity of solutions reflects a wilderness of possibilities. There are large regions without contours, where the measure of promise points in no particular direction—a clueless plateau. The lower part is boxed off, keeping the search away from the upper regions until it finally breaks through at a thin spot—the lower part is a narrow canyon of exploration. And there is a tempting contoured region that contains no solution—an oasis of false promise where a problem solver might

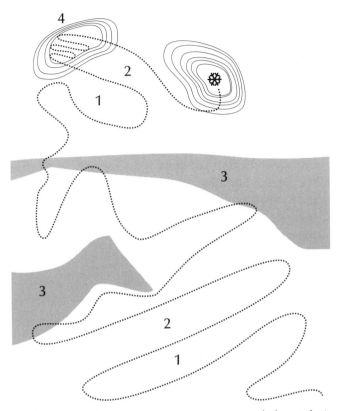

Search in a Klondike Space: 1. A large space wtih few solutions (a wilderness trap). 2. Regions with no clues pointing direction (plateau traps). 3. A barrier isolates the solution (creating a canyon trap). 4. An area of high promise but no solution (an oasis trap).

linger in hopes of finally finding a solution. Smart search in such a possibility space is a matter of thinking in ways that cope with the wilderness, plateau, canyon, and oasis traps. It's a matter of setting sequential reasoning aside and being unreasonable in a smart way, along the lines of the four operations introduced earlier: roving around flexibly, detecting hidden clues, reframing the situation, and decentering from false promise. Since Klondike spaces have their own distinctive difficulties, one might hope that the challenges of homing spaces could be left behind. But not so. Notice that Klondike spaces have small homing spaces inside them. In the Klondike figure, the small contoured region in the upper right that contains a solution is a homing space in miniature. When the agent conducting the search (for instance, a human or a computer) finally gets close enough to a solution in a Klondike space to detect promising signs, the solution still has to be extracted. Sometimes this happens reflexively, as the human mind assembles all the information spontaneously in a quick cognitive snap. Sometimes the process of extraction takes longer.

Of course, Klondike spaces and homing spaces are extreme types. Most real problems are mixed, and so are many puzzle problems. Smart search means having a sensitivity to what a problem requires and a readiness to shift between breakthrough thinking and sequential reasoning as the terrain suggests. Klondike spaces and homing spaces are worth thinking about not because they sort the world of problems into two extremes but because they anchor the two ends of a continuum, revealing the continuum more clearly.

The Structure of Breakthrough Thinking

Although wilderness, plateau, canyon, and oasis are metaphors, they have rigorous interpretations as formal features of possibility spaces. They can be described in terms of the state space of possibilities, the available operators, and the indicators of promise and success.

Wilderness of possibilities. In the language of possibility spaces, a wilderness has a large number of possible states, only a few of which are solution states. An effective search process somehow has to cope with the sheer magnitude of the state space and the rarity of solutions.

Clueless plateau. In a possibility space, a plateau is a large region of neighboring possible states where the measure of promise does not vary much, or perhaps varies erratically from state to state around an average for the whole plateau, so there is no trend. On such a plateau, a search process cannot progress from possibility to possibility with a steady improvement in the measure of promise.

Narrow canyon of exploration. In a possibility space, a canyon trap is a solutionless region of many neighboring possible states with a boundary around it that tends to trap the search process. Such a boundary can arise from the available operations. Perhaps only a very few operations from a very few states in the region take the search process outside the region. Metaphorically, there are very few paths out of the canyon. Alternatively, the boundary can arise from the measure of promise. The measure may drop to very low values all the way around the region. Although the search process can penetrate those areas in principle, it tends not to because of the low promise. Also, these two kinds of boundaries can work in combination to create a canyon.

Oasis of false promise. In a possibility space, an oasis is a state where the measure of promise has a relatively high peak that is not quite a solution state. The search process tends to circulate near the deceptive peak in hopes of finding a full solution nearby, rather than venturing off to possibilities of lower promise.

The five-phase pattern of breakthrough thinking introduced in Chapter 1 also finds an explanation in the language of possibility spaces.

1. *Long search.* Why are insights preceded by long searches? Because the space is large with few solutions (wilderness), the measure of promise does not point a clear and systematic direction (plateau), the measure of promise and available operations tend to constrain the search to limited regions (canyon), and the measure of promise yields high points with no real solution that cause the search to linger fruitlessly in their neighborhood (oasis).

2. *Little apparent progress.* Why is the apparent progress minimal for most of the search? The same reasons apply. In particular, only close to a solution does the measure of promise offer a clear guide to homing in on it.

3. *Precipitating event.* What causes precipitating events? A precipitating event can take different forms. It may simply be the arrival of the search process at a small homing subspace within the larger space. The search process then relatively quickly converges on a solution. Alternatively, the precipitating event can be some cue, internal or external, that leads the search process to escape from an oasis or canyon into another region that, relatively quickly searched, leads to a homing region and a solution.

4. *Cognitive snap.* What is the cognitive snap? The rapid homing process that occurs when the search process

finally arrives at a solution-containing homing space within the larger Klondike possibility space. The homing space allows quick convergence to a solution.

5. *Transformation.* Why the sense of transformation? Solutions tend to surprise because they often involve escape from an oasis of false promise or from a narrow canyon of exploration to a solution of a very different kind than expected.

An unusual and important feature of these explanations is that they do not specifically concern the human mind. They have little to do with a creative knack or a peculiarly flexible cerebral cortex. A Klondike possibility space is a breakthrough waiting to happen!

The search process that achieves the breakthrough may unfold in the human mind, as an artist, scientist, inventor, or businessperson explores alternative perspectives. It may occur in a computer, as automated processes of heuristic search examine a large set of possibilities. It may happen in the course of the long blind search of biological evolution, as the random shuffling of genes tries this and that new prototype for survival and reproduction. (This theme is revisited in the last chapters of the book.) Whatever the setting, most fundamentally the phenomena of breakthrough thinking derive from the underlying structure of a Klondike space.

Jack London's Klondike

One of the many outdoor tales of Jack London is a short story called "All Gold Canyon." In a few pages, London relates the struggles of a prospector to find gold. The prospector comes with the basics: a pick, a shovel, a gold pan, and most of all a hungry spirit. He chooses to start his dig in this particular canyon because there is "wood an' water an' grass an' a side-hill! A pocket hunter's delight . . . a secret pasture for prospectors and a resting place for tired burros, by damn!"

London's prospector begins with a shovelful of dirt from the edge of the stream below the side hill. He pours it into his gold pan. He partially immerses the pan in the stream and with a circular motion sluices out most of the dirt until only fine dirt and the smallest bits of gravel remain. Now comes the slow and deliberate work, the prospector washing more and more delicately until the pan seems empty of all but water. But with a quick semicircular motion that sends water flying over the rim into the stream, he reveals a thin layer of black sand on the bottom of the pan. A close look discloses a tiny gold speck. He drains more water over the black grains. A second speck of gold appears.

He pursues the painstaking process, working a small portion of the black sand at a time up the shallow rim of the pan. His efforts yield a count of seven gold specks. Not enough to keep, but enough to charge up his hopes. He continues down the stream repeating the same tedious procedure—a pan of gravel, the careful washing, the meticulous teasing out of tiny specks of gold. As he works his way downstream, his "golden herds" diminish. A pan yields one speck, another none. So he returns to where he began and starts panning upstream. His tally of gold specks mounts to thirty, then pan by pan dwindles to nothing. He has homed in on the richest point in the stream, but still nothing worth keeping. The real treasure lies above, some-where on the face of the side hill.

A few feet up from his first line of test pans he begins digging a second row of holes, crosscutting the hillside. Fill the pan, carry it to the stream, pan out the gravel, count the flecks—each tedious cycle gathers more information. He works his way up the side hill in rows of holes. The center of each row yields the richest pans, and each row ends where no gold specks appear. The rows grow shorter as he mounts the hill, forming an inverted V. The converging sides of the V mark the boundaries of the gold-bearing dirt.

The apex of the inverted V is the prospector's goal, where "Mr. Pocket" resides. As the prospector mounts the hill, the pans get rich enough for their yield to be worth saving. But the work grows harder. As the sides of the V converge, the gold retreats underground. The gold at the edge of the stream was right at the roots of the grasses. Then it lies 30 inches down, then 35. Then 4 feet, then 5.

Finally the sides of the V come together at one point. He digs his way 6 feet down into the earth. His pick grates on rotten quartz. He digs the pick in deeper, fracturing the rock with every stroke. He holds a fragment of the rotten quartz in hand and rubs away the dirt. Half the rock is virgin gold. More scrabbling about yields nuggets of pure gold. Eventually the prospector draws 400 pounds of gold from the find.

Can London's tale of the real Klondike have meaning in the esoteric conceptual world of possibility spaces and smart search? Indeed it can. London's prospector carries out a smart homing search for gold, progressing systematically up from the creek bed toward the source. Each hole he digs probes a possibility, yielding some gold and helping him to focus his search better. In Klondike terms, London's prospector searches a homing space within the larger Klondike wilderness.

Part 2

The Art of Breakthrough Thinking

In which we devote one chapter each to the four key operations of breakthrough thinking—roving, detecting, reframing, and decentering—through historical examples and puzzles.

5

Thinking's Big Bang

Roving Far and Wide

The first challenge of finding anything worthwhile in a Klondike wilderness is that there is just so much of it! Our conventional image of a breakthrough involves leaps of the imagination, a matter to be looked at more closely a little later. However, all too often there is no natural path even for imagination to follow. One just has to rove through the wilderness to see what turns up. And, if the wilderness is big, one has to rove with reasonable efficiency.

A classic example of this, discussed in Chapter 1, was Thomas Edison's draghunt technique, where he and his staff would conduct massive searches for good prospects, as in the case of a suitable filament for a lightbulb. Although Edison's large-scale searches might seem clumsy, other episodes in the history of invention testify to the viability of the approach. Edward Rosinski and a colleague developed the zeolite catalyst in the late 1950s, an innovation which increased the production of gasoline from petroleum by a huge 30 percent. To determine the optimal form of the catalyst, the team systematically tested hundreds of combinations of conditions.

Medical science benefits from systematic roving as well. Large-scale automation of the search for variations appears in the testing techniques used for probing soil samples to find bacteria that produce potentially useful antibiotics. The techniques involve assembly-line-like automation of culturing methodology. Using this procedure a team led by William Campbell discovered the antibiotic ivermectin, which prevents river blindness, a vicious parasitic infection native to Africa, as well as treat heartworm in animals and other parasitic disorders.

Matthew Plunkett and Jonathan Ellman review another contemporary methodology in the same spirit: combinatorial chemistry. Development of new drugs normally is a painstaking process that begins with a lead—a chemical with promising effects but problems of toxicity and insufficient strength—and tries to improve it through costly and time-consuming step-by-step modifications. In contrast, combinatorial chemistry explores thousands, even millions, of variations in parallel through systematic processes of mixing different combinations, using automation. Pursued by various chemical and drug companies, combinatorial chemistry has yielded drugs now undergoing clinical trials, as well as promising high-temperature superconductors and liquid crystals for flat-panel displays.

Such methods of roving are a far cry from flopping around. Sometimes people make discoveries by flopping around, but the draghunts of Edison, the examination of alternative forms of zeolite, the searches for new antibiotics, the use of combinatorial chemistry, and like examples reveal highly systematic explorations that sample large spaces of alternatives systematically, so as to cast a wide net, avoid redundancy, and not miss important possibilities.

How to Court Lady Luck

One of the disconcerting features of roving as a way of inventing is that it seems to rely on luck rather than reasoning or imagination. But the "rather than" is a mistake here. As emphasized earlier, the art of invention consists in good part in upping luck. By definition, breakthrough problems are unreasonable. Solving them always involves a measure of luck. The question is how to up luck to the point where a payoff is likely.

One of my most intriguing professional experiences was a small conference organized by my colleague Robert Weber and me to include a few cognitive psychologists, some historians of technology, and several world-class inventors. In the last category, the participants were William Campbell, head of the team who developed the antibiotic ivermectin mentioned earlier; James Hillier, one of the developers of the electron microscope; Paul Morgan, one of the developers of Kevlar, a high-strength plastic; Edward Rosinski, lead inventor of the zeolite catalyst for cracking petroleum, also mentioned earlier; James Teeri, who designed the Soil Biotron, an underground laboratory at the University of Michigan for the study of life processes in soil; Robert Wentorf, one of the developers of synthetic diamonds; and John Wild, the principal inventor of ultrasound imaging.

The inventors joined the historians and cognitive psychologists for two days of discussion, presenting papers about how they arrived at their principal discoveries. Robert Weber and I edited a book resulting from the venture, *Inventive Minds*, which includes articles from all the participants including the inventors.

One striking lesson of the event was that inventors—or

at least these inventors—were anything but your stereotypical reclusive basement tinkerers. They proved generally to be lively, witty, sociable, and broadly humanistic. Indeed, most of them worked as key figures in large industrial laboratories where people skills were a definite asset.

Another lesson concerned the long-haul character of invention. One striking feature of almost all the inventions reviewed—not only the contemporary ones but the work of the Wright brothers, Edison, and others—was the duration of the enterprise. There were no cases of instant breakthrough with prototypes on the table in a matter of weeks. Nearly every tale of invention unfolded over several years, with many false starts and dead ends. Another important feature concerned the pace of progress. While there were many moments of insight, there were no cases of inventions resulting from a single leap. Rather, the tales were long and complicated, with many insights major and minor sprinkled along the way.

Indeed, it turned out that one of the best ways to organize the different patterns of inventive practice concerned the use of luck. The inventors faced the wilderness of possibilities in various ways, which could be organized into the following spectrum of search styles:

1. *Sheer chance.* An invention not particularly sought gets discovered by an active searcher exploring widely and incidentally.

2. *Cultivated chance.* The searcher deliberately opens himself or herself to a variety of semi-random input, harvesting the occasional useful connections. This is one kind of roving.

3. *Systematized chance.* The searcher systematically surveys a sizable number of options that fall within a defined set,

seeking ones with the target characteristics. This is more systematic roving.

4. *Fair bet.* The searcher conceives and develops one or a few prototypes, relying on science and craft, with reasonable expectations that one or another will serve.

5. *Good bet.* The searcher conceives and develops a prototype from principle and experience that probably will work.

6. *Safe bet.* The searcher deduces with formal methods something that almost certainly will do the job.

All these styles of search can be found one place or another in the inventors' accounts. However, the trend lay toward the middle—categories 3, 4, and 5. Rarely did sheer chance figure. Although some of the inventors employed cultivated chance, which sometimes made a contribution, it was hardly the mainstay. At the other end of the spectrum, safe-bet reasoning through formal methods only operated to solve certain technical problems from time to time.

As one would expect, there was extensive crafting and testing of prototypes, in the spirit of styles 4 and 5. For instance, considerable work on prototypes occurred around the development of the electron microscope and of ultrasound. However, there was also a surprising amount of systematized chance—roving hard and long in a reasonably efficient way.

The Breakthrough Logic of Brainstorming

Nearly everyone has heard of brainstorming. It is a style of conceptual roving, and almost a synonym for creative exploration of ideas through systematized chance. Nearly everyone has some sense of how to go about it. One freewheels, makes up ideas and jots them down, tries to be imaginative. Brainstorming is a style of smart search for Klondike spaces. While thinking generally follows the turnpikes or side roads, brainstorming is the mind's dune buggy.

But many people are not aware that brainstorming has a specific history and a structured design that involves more than just general freewheeling. Despite the anything-goes emphasis, brainstorming has its own discipline that contributes to its effectiveness.

Alex Osborn described the technique of brainstorming in his 1953 book *Applied Imagination*. Working in business contexts, Osborn saw the need to break discussion out of ruts and to introduce more creativity into deliberations. He designed brainstorming as a group problem-solving technique to do just that. Of course, brainstorming also can be very useful in solo thinking.

Some people are surprised to learn that brainstorming has rules. Here is a version of them:

1. *No criticism.* This is the premier rule of brainstorming. During the brainstorm itself, criticism is out. Whatever's said goes on the list.

2. *Keep moving.* Don't hover to develop details. Toss in ideas and move on. Go for quantity.

3. *Piggyback.* Besides just making up ideas out of the air, take ideas already mentioned as a point of departure, extend them, add a twist.

4. *Diversify.* Try for different kinds of ideas—ideas in contrasting categories, ideas that come from different points of view.

Brainstorming makes good sense from a Klondike perspective. Most centrally, it addresses the wilderness trap. *No criticism, keep moving,* and *diversify* all promote a flexible search that samples widely the space of possibilities. However, like most techniques of breakthrough thinking, in some ways it helps with all four of the Klondike traps. As to clueless plateaus, brainstorming does not sit on any one plateau for long. The *no criticism* rule discourages discarding ideas because they do not meet the narrow constraints of familiar canyons. The *diversify* rule helps to move the search out of canyons. As to oases of false promise, *keep moving* and *diversify* encourage leaving oases behind, while *piggyback* allows for brief elaborations and extensions of attractive ideas without getting stuck on them.

Another way to appreciate the value of brainstorming invokes a distinction from computer science and psychology between depth-first and breadth-first search. Problem solvers usually follow a depth-first pattern. They think of an approach. If it seems promising, they pursue it. If they cannot make it work, then and only then do they seek another approach.

The depth-first pattern of search does not suit Klondike spaces well. Such a search is likely to neglect vast regions of the Klondike possibility space that may contain provocative solutions. In contrast, brainstorming recommends a breadth-first pattern of search, which ranges widely rather than investing deeply in any one approach. Choices can be made later.

Besides coping with the structural traps of Klondike spaces, brainstorming helps with a psychological trap. In group contexts, criticism is inhibiting. People quickly learn that others may shoot their ideas down. This establishes a conservative atmosphere, where more divergent ideas get kept in the basement rather than put out on display—not a productive way to respond to a Klondike challenge. Even in solo brainstorming, self-criticism may have an inhibiting effect. The *no criticism* rule of brainstorming helps to liberate the mind for a while.

None of this means that brainstorming is the perfect tool of breakthrough thinking. There are downsides too. For one, an interesting body of research shows that group brainstorming in its classic form does not range as widely as it might. When people work together on a brainstorm, they achieve some variety. However, if the same people spend the same time conducting solo brainstorms, a pooled collection of their ideas shows even greater variety. This finding does not mean that group brainstorming is a waste of time. A simple mixed model helps to repair the problem: Begin a group brainstorm with a solo step where each participant jots down a number of initial ideas, and then proceed in group fashion to list those ideas and more. A further benefit of group brainstorming is social, helping to build a shared sense of the problem and comfort with solutions that can be seen as owned by all since they emerged from a participative process.

A second caveat is that brainstorming just begins the work. Good brainstormers are left with an array of ideas that need to be sorted and prioritized to focus in on the most promising ones. In group settings, this embarrassment of riches often generates confusion and frustration. Tricks like asking participants to nominate their favorite ideas from the list and dropping ideas with no nominations can help to move things along.

A third caveat is that brainstorming can easily proceed

within unrecognized assumptions rather than exploring the space of possibilities freely: Reframing is needed too. And a fourth caveat recognizes that haphazard exploration may go nowhere interesting. Some breakthrough problems benefit less from free-flying exploration than from detection of hidden clues. The next chapter examines this approach.

Theme and Variations

Brainstorming and other styles of thinking in that spirit usually occur behind the scenes, with only one idea taking the stage. However, a form that makes an appearance in a number of arts openly declares its exploration of the space of possibilities—theme and variations. Of course, a theme with variations does not include all the variations the maker explored. Nonetheless, it offers a sense of the range, particularly since artists characteristically strive for great variety, to put the theme through its paces.

The musical arts offer the most familiar cases of theme and variations, perhaps because the phrase names a certain compositional structure. Well-known examples from the classical repertoire include Bach's Goldberg Variations, the middle movement of Beethoven's *Kreutzer Sonata*, and Rachmaninov's *Rhapsody on a Theme of Paganini*. Jazz improvisation foregrounds the art of variation, tossing the theme and harmonic structure from instrumentalist to instrumentalist for individual treatment by each while the others accompany.

Theme and variations figure in the literary arts as well. Consider, for example, poet Wallace Stevens' "Thirteen Ways of Looking at a Blackbird." Stevens begins by invoking a winter landscape where nothing moves but "the eye of the blackbird." He goes on to compare being of three minds with three blackbirds sitting in a tree. He expresses how the cast shadow of a flying blackbird is a mysterious mood-evoking inscription, "An indecipherable cause."

Or consider the first few lines from Archibald MacLeish's well-known "Ars Poetica," the "art of poetry." MacLeish begins by asserting that poetry should be mute "As a globed fruit." Then he says it should be dumb "As old medallions to the thumb" and offers further images for the silence and wordlessness of a poem. Stripped of metaphor, the first lines of "Ars Poetica" say something deliciously paradoxical: A poem should be mute, dumb, silent, and wordless—almost the same thing. With metaphors attached, the lines say it in a far from wordless way, evoking much the same idea four times over with different images.

Theme and variations occur from time to time in the visual arts as well. One master of the craft was the fifteenth-century Dutch painter Hieronymus Bosch. His metier was to fill scenes of heaven, hell, and earthly sin with all manner of Frankensteinian concoctions.

Bosch's most famous work is perhaps the triptych *The Garden of Earthly Delights*. The left panel portrays heaven, the right hell, and the larger middle panel the garden proper, a sumptuous park full of naked bodies cavorting in sinful ways. The panoramic scene includes well over one hundred figures. A repeated trick is to suggest the sexual act while keeping it out of sight. A man and a woman engage in foreplay inside a transparent bubble growing out of a peculiar plant. Another pair look out from a hole in a rotten fruit. A couple dance side by side, from the waist up both their

Pencil Sketch of "The Tree Man," which also appears in Bosch's painting *The Garden of Earthly Delights.*

bodies encompassed in a fruit, with an owl, symbol of evil, sitting on a branch on top. A man carries a giant mussel shell with an amorous couple's legs protruding from it.

In the panel to the right, Bosch fills his hell with chimerical creatures inflicting a variety of horrors on lustful sinners. A demon with the head of a bird, a kettle for a hat, and jugs on his feet swallows a sinner head first, while a flock of birds flies from the victim's anus. In the back-

Sketches of Monsters by Bosch

ground, a pair of giant ears lumbers across the landscape, brandishing between them a huge blade. The centerpiece of this table of horrors is the "tree-man," with a flattened egg-shaped body the rear part of which has fallen away to reveal a tavern and revelers. A pencil sketch by Bosch appears here. The tree-man's legs turn into roots that rest on two boats in a lake. Nothing is stable, nothing is logical, everything slips and flows.

Bosch did not simply make up his monstrosities on the spur of the moment. He filled many pages with his explorations of the Klondike space of bizarre creatures. One such page appears here.

Another virtuoso example of theme and variations comes from the work of the nineteenth-century Japanese woodblock master Hokusai. His series of black-and-white prints *One Hundred Views of Mount Fuji* ranges across a myriad of settings and events, but in every image the conical icon of Mount Fuji appears. How to make Mount Fuji a significant part of the print is a continuing challenge. The solution is often straightforward, a simple presence in the background. However, occasionally the solution takes other forms. One of the more interesting prints, illustrating the pinhole effect, has the image of Mount Fuji passing through a hole in the shutters to appear inverted on the paper of a *shoji* (see illustration). Other creative placements of Mount Fuji include the shadow of Mount Fuji on the waves of the sea, Mount Fuji as viewed through a spider web, and a pile of snow in the shape of Mount Fuji.

While authors and artists often combine variations on a theme within a final product, they also explore variations on a theme in order to decide which version to use. A well-known example of such a generative journey is Pablo

Picasso's extensive series of sketches for his masterwork *Guernica*. Visitors to the annex to the Prado in Madrid, where *Guernica* is housed, find with the painting a host of preliminary studies, variations on the particular images that finally took their place in Picasso's panoramic vision of the atrocities of war.

The Breakthrough Logic of Variations

We can even turn an artist's or poet's efforts into puzzles for ourselves. A piece of paper and a pencil suffice for conjuring one's own monsters in the spirit of Bosch. A poem like the following invites a similar exercise with words and rhymes:

On the Vanity of Earthly Greatness
ARTHUR GUITERMAN

The tusks that clashed in mighty brawls
Of mastodons, are billiard balls.

The sword of Charlemagne the Just
Is ferric oxide, known as rust.

The grizzly bear whose potent hug
Was feared by all, is now a rug.

Great Caesar's bust is on the shelf,
And I don't feel so well myself.

✐ Added Vanity

Guiterman's paean to the transience of might and power is a theme easily picked up. Notice the pattern of the rhymed couplets: a symbol of strength followed by its collapse into triviality. Readers are invited to compose their own couplet to insert in the middle of the poem.

I have tried this activity several times with groups. It reliably elicits a few couplets in the spirit of the poem. To make the attempt is to throw oneself into the challenge of searching a Klondike space of sparse and scattered opportunities. So how to go about the search in a way that ups luck to the point where one is likely to find a decent line? As with Edison's search or the one that yielded ivermectin, the idea is to search where the bets are good. Since each couplet amounts to a deflation from power, it makes sense to start with a search for symbols of power . . . Napoleon, the *Titanic*, the hydrogen bomb, Genghis Khan, tyrannosaurs.

To name such examples is to set out on a Klondike journey, casting about here and there with no guarantee that any one of them will serve. However, at least it is a Klondike journey in a promising region of the possibility space. With starting points like these, the next step is to seek amusing deflations. To take clues from the poem again, if tusks become something as innocuous as billiard balls, what does Napoleon become? . . . perhaps a TV special. Or tyrannosaurs become . . . perhaps Barney. Pairing a power symbol with a deflation puts us on the track of another couplet.

The Art of the Arbitrary

One of the caveats about brainstorming mentioned earlier is that it's all too easy to rove within the boundaries of unrecognized assumptions. The use of chance to break away from expectations is a well-known remedy. Innumerable episodes from the history of invention demonstrate that chance is an engine of insight. Gutenberg, Archimedes, Darwin, and more—all were surprised by what chance encounters provoked in their ready minds. This contribution of the arbitrary to insight suggests that it's a force not just to be capitalized upon when it occurs but actively cultivated. There is, so to say, an art of the arbitrary.

The art of the arbitrary is part of breakthrough thinking. Well used, it can help the problem solver cope with the wilderness trap by roving a possibility space in a far-ranging way. As a side effect, it can also bump problem solvers off oases and jump them out of canyons, carrying the search beyond limits never recognized previously. To explore some of its tools, consider another kind of puzzle, the droodle. A *droodle* is a simple line drawing suggestive of meaning but with no obvious interpretation. The accompanying figure presents a classic example of the genre. It has a standard caption, "Four elephants examining a grapefruit."

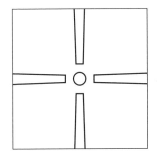

However, the figure could stand for many other things. How can we break away from the familiar solution to roam widely in the space of possibilities? Various tricks of the trade help. One good source is Edward de Bono's *Lateral Thinking*, which includes the first and third of the following tricks, among others.

Random input. The idea is simply to choose something at random and relate it to the task. One can open a book to an arbitrary page, choose a word, and see how it might connect with the matter at hand. Accordingly, I try random input on the droodle: I open a book. My eyes fall on the word *joy*. Someone might make something of that, but I don't. Then *performance*, but still nothing. Then I pick up on *urban*. I think of urban landscapes. I see the circle in the center as a moon. The elongated shapes could be smokestacks. I have a title: "Industrial Landscape: The Moon Framed by Four Smokestacks."

Tipsy triples. Three arbitrary words thrown together can create a kind of evocative pattern that can bump one out of previous ideas. I throw together three words. I don't even think about the droodle because the trio doesn't feel suggestive. I try again: *month print engine*. I don't know why these words seem evocative, but they do. Now let's see what they lead to. *Print* and *engine* remind me of printed characters. (I'm letting *month* go for the moment.) One might take the droodle as a novel character of some sort. Perhaps an asterisk but with fewer spines—an aster-less-at-risk. Interesting idea but a little obscure. Perhaps an exclamation point of some sort. Here's a title: "Four Exclamation Points Competing for the Same Dot." Here's another: "For Very Dramatic Occasions—the Quadruple Exclamation Point."

Stepping-stones. Another tool of the arbitrary is to take a step that seems strange, even unsuitable, but treat it as a stepping-stone to something else. I look at the droodle again. What could the elongated shapes possibly be? Something rather arbitrary. Well, nails say. Maybe four nails are competing for the same hole. But they have square ends, not very naillike. What has ends more square? Bolts. All right, "Four Bolts Competing for the Same Nut." Or I ask again: What could the elongated shapes be? Yardsticks. Measuring tapes. Tapeworms. That's promising. Let's see . . . "Four Tapeworms Marching in Formation down the Small Intestine."

✐ Now for Something Completely Different

With these tools of artful accident on the table, here is a droodle for readers to work on. It has the traditional title "Man with a Bow Tie Caught in an Elevator." Mentioning the usual title makes the task harder, but that is part of the challenge. What else could this figure be called? Readers are invited to use the art of the arbitrary to generate some possibilities.

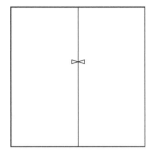

There are many other tools for the art of the arbitrary. They all show that the use of the arbitrary is far from an arbitrary process. Just because they allow people to surprise themselves does not mean that they lack logic—Klondike logic. The systematic use of accident to up luck stands worlds apart from plain old lucking out, welcome as that is when it happens.

6

Clues for the Clueless

The Dog Did Nothing
in the Nighttime

A classic moment from Sir Arthur Conan Doyle's Sherlock Holmes stories occurs when the perspicacious Holmes finds a clue in a nonevent. In "Silver Blaze," Holmes relates to Watson how he was set on the right track by "the curious incident of the dog in the nighttime." Watson complains, "The dog did nothing in the nighttime." To which Holmes responds, "That was the curious incident."

Holmes saw a clue where there appeared to be none. His acute observation speaks to one of the fundamental challenges of breakthrough thinking—cluelessness, or in our Klondike lexicon the plateau trap. Characteristically, breakthrough problems do not offer clues that lead systematically to a solution. Problem solvers find themselves on a clueless plateau, not knowing where to go.

But detection is Holmes's game. He looks harder and smarter for clues, finding significance in the insignificant, meaning in the dog's doing nothing. Holmes's style teaches a fundamental lesson about the art of breakthrough thinking. We tend to identify breakthrough thinking with lateral shifts of viewpoint, but it's worth remembering that another shift of viewpoint is simply to look more deeply. Transformation can come from more assiduous attention to the available information and sensitive flexible examination of its implications. Detecting clues where there appear to be none—clues for the clueless—is a fundamental and powerful strategy of Klondike logic.

Here is a puzzle that illustrates the point:

🖉 Ten Matches

You have ten matches to work with (or you can make a sketch). The challenge is to arrange them into two squares of different sizes, using all the matches. No breaking of matches is allowed.

The interesting thing about this problem is that it can be approached in two very different ways—freewheeling exploration and detection of hidden clues. The freewheeling approach is a matter of trying different layouts of matches. At first, nothing seems to work. No clues are apparent. Trying different combinations where the squares connect seems like a good idea. More exploration ensues. Suddenly a solution falls into place: a large square with two matches per side, the two remaining matches making a small square in one corner. It's now clear what one barrier was. There's a tendency to suppose at first that the two squares are separate rather than overlapping, a narrow canyon of exploration in addition to the clueless plateau.

However, there is another very different approach, clues for the clueless. The key clue is that the squares must be of different sizes. The problem solver might think as follows:

If there are two squares of different sizes, there must be one square with two matches per side and one with one match per side, because the next bigger size of square, three matches per side, would use up more matches than I have. Therefore there must be one square with two matches per side, using up eight. That leaves two matches. What am I going to do with them? To make another square, I have to build it onto what I already have, and the only place to do that is in a corner of the big square.

Einstein as Sherlock Holmes

Sherlock Holmes, although a fictional character, looked to the real world for clues. Albert Einstein, a very real individual, often looked to the world of imagination. In 1895 at the age of sixteen, Einstein conducted an entirely imaginary experiment that eventually was to change our conception of physical reality.

In German, such investigations are called *Gedanken* (thought) experiments. The precocious Einstein's *Gedanken* experiment involved traveling at the speed of light, alongside a light wave. He asked himself what such an experience would be like and what physical sense it would make. This fantasy journey was a clever approach to thinking about how a light wave works. Notice how sensible such a strategy would be in the case of water waves. If an investigator wanted to see the shape of a water wave more clearly, the investigator would do well to travel at the same rate and in the same direction as the wave.

Einstein's fantasy journey was a flexible leap of the imagination. However, what Einstein found at the end of it was not an answer, but only a clue. Just as Holmes recognized the silence of the dog in the night as an anomaly, Einstein recognized the imagined light wave in suspended motion beside him as an anomaly. It was an impossible situation. It couldn't happen. Here are Einstein's own words from his 1946 (published 1949) *Autobiographical Notes*:

> If I pursue a beam of light with the velocity c (velocity of light in a vacuum), I should observe such a beam of light as a spatially oscillatory electromagnetic field at rest. However, there seems to be no such thing, whether on the basis of experience or according to Maxwell's equations. From the very beginning it appeared to me intuitively clear that, judged from the standpoint of such an observer, everything would have to happen according to the same laws as for an observer who, relative to the earth, was at rest.

What clue did Einstein see in his *Gedanken* experiment that the Watsons of his day might not have? The science of the times said that just as water waves travel through water and sound waves through air, light requires a medium too, a hypothetical medium called the ether. Presumably, by travel-ing fast enough relative to the ether, one could cruise along-side a light wave and observe it suspended. But existence of the ether seemed something of an ad hoc hypothesis, since unlike water or air it did not show itself physically—unrea-sonable in Einstein's view. As a matter of symmetry, and by analogy to Newton's classic laws of motion, Einstein felt that the basic laws of physics should be the same in different frames of reference that were moving at constant velocities relative to one another. Maxwell's equations should apply to any such frame of reference. Yet they disallowed frozen light waves. Something was deeply wrong.

In his autobiography, Einstein identified in this 1895 thought experiment the germ of the theory of relativity. But only the germ, only a clue. Indeed, it was not until

1905, ten years later, that Einstein followed the trail to its climax in the special theory of relativity, a story to be concluded in the next chapter.

When Computers Think Better than People

Margaret Boden in her *The Creative Mind* points out how sometimes a human can be more clueless than a computer. A classic theorem of geometry says that the base angles of an isosceles triangle (a triangle with two sides of equal length) are equal. One elegant way to prove this is to show that such a triangle is *self-isomorphic*—triangle *ABC* is isomorphic to (has the same shape and size as) itself when flipped over its vertical axis, triangle *CBA*.

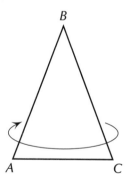

The clue is there to be seen by all. It lies in the very symmetry of the figure, an invitation to try the left-right flip. Unfortunately, this simple maneuver does not come easily to the human mind. When people solve such geometry problems, they take visual intuition as their guide. They look for relationships within the given figure rather than a transformation of the whole figure. They make constructions, typically dropping a perpendicular from the apex to the base to divide the triangle in half and relate the two sides to one another. They tend not to think in terms of a left-right reversal of the figure to create a self-isomorphism. From the human perspective, this particular solution is unreasonable—hard to get to by reasoning. Boden points out that Euclid himself, the father of Euclidean geometry, missed the idea. Pappus of Alexandria came up with the self-isomorphism proof six centuries later.

An artificial intelligence program discussed by Boden easily discovered the proof by self-isomorphism. Why was the program's "eye" more alert than the human eye? Because it lacked human biases. For the program, finding this solution was reasonable. Self-isomorphism was nothing special or odd but just one more mapping to consider. Also, the program only used constructions as a last resort and so did not get distracted by them. Boden argues that the program should not be considered particularly creative because it had no barrier to overcome.

All this has two morals. The first is simply that clues and cluelessness are in the eye of the beholder. One sees what one is ready to see. If that were the end of the story, it would leave us all stuck with our own entrenched natures. Some of us may be Holmeses, but most of us are Watsons. However, there is a second and much more hopeful moral: What one sees depends on whether and how carefully one looks, whether one puts oneself in a position to see. Einstein put himself in a position to see what he was ready to see by performing his *Gedanken* experiment. Pappus found the symmetry that Euclid missed. And most anyone can put himself in the position to see the logic of the Ten Matches problem. It's just arithmetic. Once you start on that path, the problem unravels and reveals itself.

When Imagination Signifies Nothing

Mathematical and scientific problems often lend themselves to clues for the clueless—finding subtle clues and building on them. Can the same be said for more everyday problems? Often, it can. A nice example comes from a problem about anchoring sailboats.

A number of years ago, I taught a course in which the students systematically tried out a number of problem-solving techniques. One week we focused on *Synectics*, a well-known approach to group creative problem solving that emphasizes the use of metaphors and analogies. Synectics is a style of brainstorming. The techniques of Synectics and brainstorming in general can be very generative, yielding surprising connections and creative solutions.

However, in this case we discovered that the use of metaphors and analogies is no substitute for seeing the underlying logic of a situation. Powerful as it can be, imaginative brainstorming (which will be explored more in a later chapter) sometimes suits all too well the classic words of Shakespeare from *Macbeth*, where Macbeth in his frustration and despair cries out that life is "A tale full of sound and fury, signifying nothing."

When does imagination miss the mark? When it neglects to examine closely the fundamental logic of the situation for clues. In this particular case, one of the students who enjoyed sailing volunteered a sample problem on which to try out Synectics. It ran as follows.

✐ The Drag Alarm

When we anchor at night, sometimes currents or the wind cause the anchor to drag. There's a risk of dragging far enough so we hit rocks or shoals. The usual answer to this problem is to stand watch. But it's a real pain, because we're all tired by that point. It would be great if there were some kind of alarm that would ring when the anchor dragged. Then we could sleep, and we'd wake up if there was a problem.

As usual, readers may want to think about the problem before continuing.

The group addressed the problem of the drag alarm. There was a healthy brainstorm of ideas, applying the techniques of Synectics. The participants generated a rich range of analogies. However, gradually a dilemma emerged. None of the ideas seemed to be practical, although they certainly showed imagination.

I decided to try a different approach altogether, looking for an underlying logic to the situation. For an alarm to ring, the alarm mechanism had to "know" that the boat was slipping. To know the boat was slipping, the alarm had to have a point of reference external to the boat. What could that point of reference be? Not the shore, because detecting the shore at a distance would require radar or automatic siting devices or something equally exotic. It had to be something ready at hand—the bottom under the boat. How could the alarm know where the bottom was? By a physical connection of some sort, a line like the anchor line that reached to the bottom.

This led to the idea of an "alarm anchor." The plan was to place a second, lighter anchor overboard, leaving the line a little loose and attaching it to something noisy like a pile of cans. If the main anchor slipped, the drifting boat would bring up tension on the alarm anchor's line and topple the cans, waking up the crew.

Not a boatman, I have no idea how practical this is. But

it is a step in the right direction. The point is that sometimes freewheeling imagination can function cluelessly. It can pass entirely by crucial problem constraints that have not been recognized. In the group's initial approach to The Drag Alarm, all the brainstorming neglected a point with crucial implications: To detect the dragging anchor, the alarm had to have information from a fixed point of reference.

The Importance of Pushing the Reasoning

It's easy to speak of detecting subtle clues and following their implications, but of course not so easy to do. A case study illustrates the point.

✐ The Clock
One day a mantel clock that chimed the hours and quarter hours (one chime each quarter hour) struck twenty-seven times within the span of an hour and one minute. Yet there was nothing wrong with the clock, and all this happened in a natural and appropriate way. How could that be?

Readers may want to pause to work on the problem before continuing. This unreasonable problem involves several Klondike traps. Hints appear along the way in the following discussion, and ultimately the solution comes through a clue.

I have often investigated breakthrough thinking through informal clinical interviews. Recently I explored some insight problems with a friend I'll call Emilia. I began by giving her some other puzzles, which she solved readily.

Emilia is bright and has an intuitive flare for Klondike logic.

However, when I presented The Clock to Emilia, she immediately fell into a combination oasis of false promise and narrow canyon of exploration. Figuring with the numbers, she noted that eleven chimes and twelve chimes plus the quarter hours between 11 and 12 yield twenty-six chimes in all. Only one more ring would make the twenty-seven. This near-solution was the oasis of false promise. She concluded that somehow she needed to introduce another quarter hour into the scenario. Sensible as it seemed, this assumption created a narrow canyon of search that led nowhere.

After a few minutes, Emilia was ready for a hint about breakthrough thinking. "You're in a canyon," I said, "And you can't see the sides. You're making an assumption or taking something for granted that you shouldn't."

A methodical thinker, she explained my advice back to me with precision. Then she asked for a restatement of the problem, which I provided, to try to discover whether anything had slipped by her. Then she began to explore her unrecognized assumptions. "Was more than one clock involved?" she inquired, a good question, which I answered negatively. She considered hours not in the neighborhood of 11 or 12, to see whether something might be made of that. She said she would try to work backward. Originally she had been thinking of eleven strikes on the hour at 11 o'clock, followed an hour later by twelve strikes. Now she would start by imagining the clock striking 12 and try to tell the rest of the story around that.

This might have worked. But it didn't. She concluded that she would still need the eleven strikes from 11 o'clock. Emilia tried several other moves to bypass or redefine boundaries. Emilia was doing the right thing in general. Unfortunately, she did not happen to hit on the right idea. Emilia was ready for another hint.

This time I warned Emilia about the oasis. "So," Emilia replied, again signaling her understanding, "I have to watch out for a tendency to make a tempting near-solution work instead of abandoning it. A similar problem might help." She mentioned The Nine Dots problem, which she already knew, and took inspiration from it. She vowed not to stay with the 11 and 12 but find another solution, an effort to avoid the oasis. "Suppose the time doesn't matter," she proposed. But she could not see how that would work out. "How about echoes?" she asked. I informed her that echoes were a clever idea but not the official solution. She asked for another hint.

Since Emilia found herself on a clueless plateau, I encouraged her to challenge cluelessness and look harder for clues. "Reason from the logical necessities of the situation," I said. "Look at your givens and think about what they *must* imply."

I had it in mind that to get to twenty-seven chimes in one hour and a minute, the clock *must* sound an extra time either at the quarter hour or on the hour. For instance, it might strike 12 twice. Under what conditions could this conceivably happen? However, this line of reasoning escaped Emilia. Instead, she wondered whether the person hearing the clock somehow could have been mistaken. I told her that the person heard correctly. She wondered whether a clock somehow could have twenty-seven chimes, but dismissed that. She drew a diagram, but it yielded nothing new.

I offered one more general strategic hint. I noted that she had mentioned some conclusions that seemed absurd.

Maybe she should focus on those. "One idea is this," she said. "It chimes two 12s instead of 11 and 12. But that can't apply because there's nothing wrong with the clock." In other words, she specifically formulated and rejected the key hypothesis.

After some more mulling, again she found herself stuck. I offered a very specific hint this time. I said, "You mentioned that the clock couldn't strike 12 twice."

"So somehow it can," she replied. "Somehow it can." Within a few seconds she had the solution. The clock will chime 12 two times within an hour if the owner sets the clock back one hour just after midnight on the night people adjust their clocks back from daylight savings time to standard time.

An important moral follows from this exploration of The Clock. We can do the needed reasoning if only we find the right starting point. Emilia quickly reasoned out how the clock could strike 12 twice once she focused on the possibility rather than dismissing it. The trick to breakthrough thinking based on clues is to follow what seems logically required, even though it does not appear at first to make sense. Another bit of Holmesiana, from Arthur Conan Doyle's "The Sign of the Four," expresses the point aptly: When you have eliminated the impossible, whatever remains, however improbable, must be the truth.

Two Only Seemingly Clueless Problems

As the above cases show, sometimes a plateau trap is an illusion. A more careful examination of the problem discloses clues that point the way to a solution. The path of reasoning may not only get one away from a clueless plateau but allow one to avoid narrow canyons of exploration and oases of false promise by reasoning past them. Here are two more puzzles where this kind of thinking can be helpful:

✐ Four Chains

Greta has four short chains of just three links each. She wants them reworked into one continuous circular chain of twelve links. Her jeweler will do the job for three cents to open a link and two cents to close it. Greta figures out a way to get the job done for fifteen cents. What does she tell the jeweler to do?

✐ The Four Trees

A landscape architect has a passion for symmetry. The designer decides to place four golden rain trees in a park so that each stands exactly the same distance from all the others. How does the landscape architect position the trees?

These problems have the same character as the Ten Matches puzzle: They can be approached through freewheeling exploration but they lend themselves to clues for the clueless as well.

The Solutions

The Four Chains. This problem invites imagining different ways of unlinking and relinking the chains. The obvious but insufficient solution involves opening a link at the end of each chain, for four links and a cost of twenty cents. For fifteen cents, there must be a better way.

Freewheeling exploration eventually may lead to the idea of taking one of the chains and opening all three links. Those links can connect the remaining chains in a loop, for a total cost of fifteen cents. However, a plateau trap of no obvious clues and a canyon trap—the assumption many people make that the same operation should be performed on each chain—stand in the way. Here is the clues approach. Fifteen cents means that Greta told the jeweler to open and close just three links. Which three links? There is no point in taking one link each from three of the original chains, because this strategy still leaves four pieces to be connected without enough links to do the job. Therefore, the freed links must all come from a single chain.

__The Four Trees.__ This problem invites exploration of various layouts, typically three trees in a triangle with one in the middle. That might seem to be a solution, but the tree in the middle is a different distance from the others than they are from one another. Then how else could one arrange the trees? Perhaps a square, but that does not place the trees equidistant from one another either. What about putting them all in a row? That does not meet the conditions. Eventually, the canyon becomes clear—the assumption that the trees lie on a plane. Three of the golden rain trees can rest at the corners of a triangle and one on a steep hill in the middle.

However, one can get to a solution with much less search by reading the clues and reasoning from there. To place three trees equidistant (never minding the fourth for a moment), there must be an equilateral triangle—the only shape that positions three points equidistant from one another. The fourth tree, to be equidistant from the others, also has to make an equilateral triangle with each pair of them, so it has to be off the plane—on

a hill or in a pit. As to the pit, drainage would be some-
thing of a problem, and the landscape architect certainly
does not want a golden rain tree in a hole in any case.

Creative Cluelessness

When we're looking hard for the subtle clues, sometimes
the clues we already see can get in the way. There are
moments where it's worth deliberately putting oneself into
a state of cluelessness, to see what new kinds of clues then
show up. In the hit film *Field of Dreams*, Kevin Costner's
character becomes obsessed with building a baseball sta-
dium in the middle of nowhere. He has a vision that "they
will come": The players and fans somehow will find their
way there. We cannot expect visions to guide our inventive
lives with any regularity, but the idea of starting in the
middle of an arbitrary field has its value.

Ronald Finke, a creativity researcher at Texas A & M
University, offers a provocative guide to this with the concept of *preinventive forms*. Preinventive forms are tools for
starting in the middle of an almost clueless nowhere and
making something of it. Finke and his Texas A & M colleagues Tom Ward and Steven Smith place the idea of
preinventive forms in the context of an overall theory of
creativity in their 1992 book *Creative Cognition*. They
introduce the *Geneplore* model of creativity, referring to the
generative and exploratory cognitive processes typical of
how creativity happens. The Geneplore model identifies
two stages of creativity, the generative stage and the
exploratory stage. During the generative stage, the creator
generates preinventive structures such as visual patterns and
mental blends by memory retrieval, association, synthesis,

analogy, and other acts of mind, guided by a sense of novelty, ambiguity, incongruity, divergence, and potential meaningfulness. During the exploratory stage, the creator explores and seeks to interpret the preinventive forms to see what creative possibilities they afford, leading to modifications of existing structures or production of entirely new structures. This cycle often repeats several times.

The best way to communicate the concept is to do an exercise with it, one that mirrors the kinds of experiments Finke has conducted.

✐ Gizmo

Step 1: Take a look at the various shapes illustrated. Pick any three of them, and assemble them mentally into a form that looks as though it might be good for something. Do not devise a specific invention, just something that seems vaguely as though it would have a purpose. Take just one minute for this step.

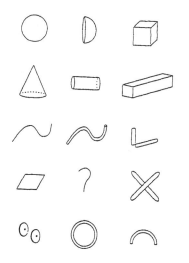

Step 2: Arbitrarily—not calculatedly—choose one category from the following list of eight:

Furniture   Personal items
Transportation   Appliances
Scientific Instruments   Tools and utensils
Toys and games   Weapons

Step 3: Now try to interpret the form devised above as some gizmo in this category—a piece of furniture, a personal item, a device related to transportation, or whatever. Again take only one minute.

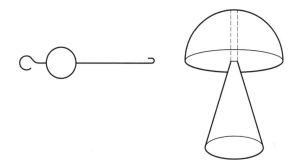

In his studies of people thinking according to this pattern, Finke found that people often came up with interesting ideas this way. The above figure displays two examples from Finke's studies. The form on the left, viewed as a tool, became a *contact lens remover*. The one on the right, viewed as a personal item, became a *universal reacher* for retrieving objects in hard-to-reach places.

Getting responses from a number of participants, Finke rated the inventions on both originality and practicality. He classified inventions as creative when they were both original and practical, or simply practical when they seemed functional but lacked originality. Finke came to the following conclusions:

- People often arrive at creative inventions and practical inventions despite the arbitrary character of the task and the minimal time allotted. Out of 360 trials there came 120 practical and 65 creative inventions.

- Creative inventions proved less likely when a participant was given the category (furniture, personal items, etc.) before rather than after choosing the parts to fashion the preinventive form. Apparently this reduces the divergent tension between the form and the category.

- Creative inventions proved less likely when a participant, after constructing a preinventive form, could choose the category rather than being given it. Again, apparently this reduced the divergent tension between form and category.

- Finally, creative inventions proved less likely when people worked with preinventive forms others had fashioned rather than ones they themselves had assembled. Apparently people have a real sense of what looks like a potentially meaningful form to them.

This intriguing study gives empirical testimony to the magnetism of expectations in limiting creative thought. Creativity tended to suffer when people received or chose the clue of a category in advance. This hardly means that one should routinely ignore the constraints of the problem at hand. However, it does emphasize that sometimes there is no problem or none very well demarcated. Then, the field of dreams of a preinventive form may be the best approach to something interesting.

7

Walking through Walls

Why Did the Chicken Cross the Road?

Most people have encountered this antediluvian riddle and its answer. Why did the chicken cross the road? To get to the other side. Indeed, the official answer can be considered a kind of antihumor. It's at least a little funny because it refuses to offer some splendid startling switcheroo and offers instead the most straightforward possible answer—itself a kind of switcheroo. But of course, it's not *that* funny.

Now what's the real reason the chicken crossed the road? Well, that depends on how you look at the chicken, the road, and the rest of the universe. A few months ago on the Internet, a list circulated explaining how a number of notables from history might answer the chicken riddle. For instance, there are religious and philosophical answers such as . . .

BUDDHA: If you ask this question, you deny your own chicken nature.

MACHIAVELLI: The point is that the chicken crossed the road. Who cares why? The ends of crossing the road justify whatever motive there was.

RALPH WALDO EMERSON: It didn't cross the road; it transcended it.

JEAN-PAUL SARTRE: In order to act in good faith and be true to itself, the chicken found it necessary to cross the road.

There are also scientific answers . . .

DARWIN: Chickens, over great periods of time, have been naturally selected in such a way that they are now genetically dispositioned to cross roads.

B. F. SKINNER: Because the external influences, which had pervaded its sensorium from birth, had caused it to develop in such a fashion that it would tend to cross roads, even while believing these actions to be of its own free will.

CARL JUNG: The confluence of events in the cultural gestalt necessitated that individual chickens cross roads at this historical juncture, and, therefore, synchronicitously brought such occurrences into being.

ALBERT EINSTEIN: Whether the chicken crossed the road or the road crossed the chicken depends upon your frame of reference.

All these answers remind us that there is no one chicken reality, no one answer to the chicken question. As with so much else in the world, it depends on framing—perspective, frame of reference, point of view, background assumptions. And that is crucial to breakthrough thinking, whether about chickens or the laws of the physical universe.

Einstein on the MTA

In his autobiography, Einstein identified his 1895 thought experiment as containing the germ of the theory of relativity. But he lived with that germ a long time. It was not until 1905 that he achieved another breakthrough that put it all together. In a famous paper entitled "On the Electrodynamics of Moving Bodies," Einstein advanced two axioms to form a new paradigm of physics. The first stated flatly that the laws of physics should be the same in any frame of reference moving at a constant velocity, that is, not accelerating. The second stated flatly that the speed of light was the same in any such frame of reference. Not only could you not travel fast enough to suspend the light wave, Einstein had concluded. You could not even begin to catch up with it. Light would always be moving at speed c relative to your own frame of reference.

While the first axiom may seem a sensible symmetry, the second is peculiar indeed. A well-known folk song of many years back tells the story of poor Charlie, who got trapped aboard what was then called the Boston MTA (Metropolitan Transit Authority subway system) because of a fare increase that required riders to pay the extra as they left the train. Charlie did not have the money to get off, so his wife had to come down to the Scollay Square station "every day at quarter past two," to hand Charlie a sandwich through the train's window as "the train comes rumbling through." Suppose Charlie's train happens to be an express and it doesn't even stop at the Scollay Square station. The train rolls through at 25 miles per hour. To pass Charlie a sandwich through the window, his wife runs alongside the train at 10 miles an hour. Relative to his wife, the train only moves at 15 miles an hour and it's an easy handoff.

The relativistic version is not so forgiving. Imagine that the fare increase traps Charlie on a light-train, a subway system that converts its passengers into light for truly rapid transit. The light-train lasers through the Scollay Square station at the speed of light, about 186,000 miles per second. Determined to pass Charlie his sandwich, Charlie's wife runs along beside the tracks at a truly extraordinary 10,000 miles per second. However, she is startled to note that the light-train still zips by her at 186,000 miles per second. At noon the next day, Charlie's wife gets out her track shoes and runs alongside the light-train at a full 100,000 miles per second. Amazingly, she finds the train and Charlie still passing by her at 186,000 miles per second. No sandwich for Charlie.

How can such seemingly simple arithmetic about velocities possibly fail? How could it be that one could not slow a light wave down relative to oneself by traveling along beside it fast enough? This question lay at the core of Einstein's 1905 breakthrough. According to Einstein's account, the fundamental issues had been brewing in his mind for some time. But one morning in the spring of 1905, upon waking up, Einstein found it all coming together. It appears that the key step was to challenge a fundamental and simple premise we all take for granted: the constancy of time.

Let's turn back to the relativistic Charlie example. If we presume that Charlie's time and his wife's time are measured on the same absolute clock that applies anywhere in space and at any velocity, Charlie's wife should be able to run alongside the light-train fast enough to pass Charlie a sandwich. So, Einstein reasoned, there must be something wrong with the assumption. Suppose instead that time itself is not invariant, that the clock in the station and the watch carried by Charlie's wife do not stay in synchronization. Then the conflict does not arise. The light-train could pass Charlie's wife at 186,000 miles per second, even though she is running alongside it at 100,000 miles per second, because time is running slower for Charlie's wife than for an observer standing still in the station. If Charlie's wife checks her watch against the clock in the station after her run, she will find that it has lost time relative to the station clock. She has spent less time running than has passed in the station!

It may seem odd that Einstein would substitute such an exotic idea (backed up with the appropriate mathematics) for the simple addition and subtraction of velocities, but Einstein was driven by a vision of fundamental simplicity in physics. Suspended light waves did not make sense, so they had to go. Other related phenomena posed similar problems. To resolve those anomalies, something else had to give—and the something turned out to be the constancy of time.

In terms of Klondike theory, Einstein escaped from a canyon, the presumed constancy of time. The canyon is a very fundamental one, a tacit assumption reinforced by everyday perception, as Einstein explicitly recognized. Yet Einstein's efforts to resolve the paradoxes he saw in the physics of his day finally led him to question the usual frame, where time was constant, and to make up another, where it wasn't.

Reframing Problems

Seeking new ways of looking at problems is a fundamental response to narrow canyons of exploration, and to some extent the other Klondike dilemmas. If the way you've been coding or representing the situation to yourself is not going anywhere, perhaps the problem is unreasonable—so why not try something different? Einstein accomplished exactly this by imagining what it would be like to ride along with a light wave, adopting a fresh viewpoint with startling implications. He did it again ten years later when he reframed time itself as something that could vary locally.

Many scholarly and popular authors interested in insight have emphasized the importance of reframing. The Gestalt psychologists recognized the hegemony of pattern and the role of pattern breaking in good thinking. Edward de Bono, well known for his practical thinking guides, underscores how familiar categories get "stamped in," standing in the way of more flexible thinking.

Cognitive psychologist Stellan Ohlsson, writing about insight, emphasizes how fundamental arriving at a new representation of a problem often is. It can bring the solution suddenly within the "horizon of mental look-ahead." Able to see how the new representation leads to a likely solution, the problem solver has a sense of insight. Ohlsson identifies three different conditions that cue the problem solver to stand back and consider reframing a problem:

1. *Restructure when stuck.* Having no place to turn within the current representation recommends seeking a new one.

2. *Restructure upon novelty.* When interesting and surprising aspects of the problem emerge, this is a natural time to seek a new representation that accommodates them.

3. *Restructure upon overload.* When it is difficult to keep track of all the information and trials at hand, a new representation that simplifies and organizes the situation may help.

The following simple game demonstrates how dramatically reframing the way a problem is seen can reduce the difficulty of a problem.

✐ The Digits Game

Two opponents find on a table the digits 1 through 9 written on slips of paper. Each player alternates in picking up one slip of paper. The goal is to acquire any three numbers that sum to 15 before the opponent does so. The first player begins by selecting the 2. What is the best move in response?

Unfamiliar though this game may seem, most people already have all the knowledge they need to choose the best response. In fact, most people have enough knowledge to play the entire game with perfect strategy. But connecting that knowledge to The Digits Game requires reframing the situation in a surprising way.

I recommend that readers not struggle with this problem, because the solution calls for a fresh representation not likely to be familiar to most people. There are arrays of numbers called magic squares, where adding the numbers in any row or any column or along the diagonals yields the same sum. One magic square includes the digits 1 through 9 arranged as shown here. With this clue, the chances of figuring out how to play the Digits Game are somewhat better.

2	7	6
9	5	1
4	3	8

The secret is this. The Digits Game is an analog of tic-tac-toe. Picking a number is like putting an X or an O in a box. The rows, columns, and diagonals correspond to situations where three numbers sum to 15. For instance, if an opponent has started by choosing 2, that means the opponent has made a corner move. As in tic-tac-toe, you have to block right away to prevent the opponent from forcing a win on subsequent plays. Good tic-tac-toe strategy says that you must choose the center square to avoid a trap, picking up the 5. The Digits Game is a wonderful example of how radically a new frame of reference can transform a problem.

The Digit Game also reminds us that analogies are one of the most powerful tools of reframing, although by no means the only one. The episodes of Archimedes in the bath, Gutenberg at the wine festival, and Darwin reading Malthus all involved analogies. While these analogies have considerable stretch to them, linking contexts rather remote from one another, many generative analogies are much closer to their subject. McGill professor Kevin Dunbar has done extensive case studies of research in molecular biology laboratories to examine the moment-to-moment evolution of ideas. Dunbar's studies show that analogies figure commonly in the development of theories, experiments, and explanations. However, in this setting, the analogies are typically close. Dunbar sorted analogies that occurred into three categories—within the organism, across organisms, and nonbiological—and found that the first two accounted for almost all of them. When nonbiological analogies occurred, they always served to explain ideas, not to generate theories.

The Copernican Revolution as Reframing

New representations are helpful not just for puzzles but for mysteries on a grand scale. A case in point is one of the fundamental scientific discoveries of all time, the Copernican revolution. The problem to be solved concerned the march of the heavens under which the human race has lived for millions of years. From before the time of Christ, certain stars stood accused of disorderly conduct. They moved against the pattern of the rest of the stars, shifting in one direction but sometimes reversing direction for a while as month followed month and year followed year. These were what we today recognize as the planets. Indeed, the term *planet* comes from the Greek word *planasthai*, meaning "to wander"—just what these vagrant stars did.

How to explain the odd paths of these wanderers against the fixed stars became one of the key questions of early science. Until the middle of the second millennium, a simple representation dominated thinking about this problem, one derived from Aristotle and the Catholic church. The Earth stood at the center of the universe. The planets rotated slowly around the Earth, their strictly circular motion expressing nature's most perfect form. However, accounting for the occasional retrograde motion of the planets proved to be a challenge. Where did those peculiar reversals of direction come from?

The Egyptian astronomer Ptolemy, working in the second century A.D., offered an elegant solution to this puzzle: Not only did the planets rotate around the Earth, but they also made smaller circles in the sky called epicycles. A fancy kind of carnival ride helps to explain how Ptolemy's solution worked. This ride has a long boom that rotates. On each end of the boom is a small Ferris wheel. If you take the ride, you find yourself spinning around in one of the small Ferris wheels, with the whole thing spinning around the central hub. Now I'd rather not take such a ride, so imagine I'm watching a friend as he swings to the very top. Which way is he moving? If the main boom rotates clockwise, he might be moving to the right, like the hands of a clock at noon. But if the little Ferris wheel he's on spins counterclockwise fast enough, he'll be moving to the left at the top, despite the clockwise motion of the main boom. Likewise, by choosing the right rates of rotation and diameters of wheels, Ptolemy could generate a reasonable model of the occasional retrograde motion of the planets against the stars.

However, Ptolemaic predictions had their inaccuracies. In the years after 1500, the Polish astronomer Copernicus, after extensive data gathering, defended a radically different model, a fundamental change of representation. The planets rotated not around the Earth but around the Sun—as did the Earth. The odd motion of a planet against the fixed stars reflected the combined effect not of a little wheel attached to a big wheel, but of the Earth's orbital motion along with the planet's orbital motion, no epicycles needed. While Copernicus's circular motion did not deliver sufficiently accurate predictions, close upon the heels of Copernicus came Johannes Kepler, who worked out a model where the planets, including the Earth, moved in elliptical rather than strictly spherical orbits—the model that survives today. Although the story of the Copernican revolution is far more complex than these few paragraphs say, at its heart lay two fundamental shifts of representation—from a geocentric to a heliocentric conception of circular motion, and then from circular to elliptical motion.

Three Problems of Reframing

Here are three more problems that invite solutions by finding a new representation, although one might approach them in other ways as well. Unlike The Digits Game, none of them requires specialized knowledge.

✐ Giraffes and Ostriches

There is a peculiar neurological disorder where the sufferer can see parts but not wholes. A victim of this disorder visited a section of a zoo one day and, among other things, saw thirty eyes and forty-four legs. A companion explained that the two were viewing giraffes and ostriches. "Aha!" said the sufferer, whose reasoning was not at all impaired. "I know exactly how many giraffes and how many ostriches I saw." So how many were there? (As a point of background, there are indeed neurological disorders where people cannot integrate individual features to recognize a whole, although this puzzle takes this incapacity to an extreme.)

✐ The Monk's Journey

One day at sunrise, a monk started up a mountain, aiming to fast and meditate overnight. He eventually made it to the top and spent a solitary night in deep and reverent contemplation. The next morning exactly at sunrise, he started down the mountain by exactly the same path, but not at the same pace, sometimes walking faster in the downhill direction but sometimes strolling slowly to appreciate nature. On the way down, he recognized many familiar places. He fell to wondering about this question: "Was I ever at the very same place at the same time of day coming up as I am now going down?" Must there have been a point on his trip down where he was at the same place at the same time as coming up or not, and why?

✎ The Four Beetles

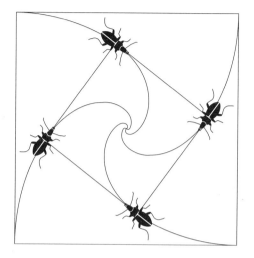

Four beetles trained to follow one another rest at the four corners of a square measuring 1 foot on each side. At the word "go," the trained beetles start to crawl toward one another. Each one crawls and crawls, always aiming at the beetle in front of him. This causes each beetle to turn gradually rightward toward the center, because the beetle in front is traveling to the right and the beetle behind must turn rightward to keep pointed at the beetle in front (see illustration). The result of this is that the beetles quickly spiral in toward the center of the square and meet. How long is the path that each one of them traces? Although the curve looks complex, this problem requires no calculus nor even algebra.

Solutions by Reframing

Each of these three puzzles yields to reframing.

Giraffes and Ostriches. The first problem is the most

accessible. With algebra, it can be solved in a thoroughly routine way. One can treat the problem as a homing problem that might appear in any algebra text, set up two simultaneous equations, and solve them.

However, a cleverly selected representation makes all the algebra unnecessary. One might think in terms of the fronts of giraffes and ostriches, neglecting the back legs for a moment. The thirty eyes mean one can build fifteen fronts, using up thirty of the legs in the process. The fourteen legs left over are enough to build seven giraffes by providing seven of the fronts with rear legs as well. The other eight fronts will have to remain ostriches.

The Monk's Journey. *One of the most powerful and general strategies for insight problems as well as many other kinds of problems is to make and manipulate mental movies—not just still images but sequences of events that play out in the cinema of the mind. According to recent work by Harvard psychologist Steven Kosslyn and others, there is a cinema of the mind in a fairly literal sense. Recent research shows that mental imagery gets projected in an area of the brain that is also activated during sight. Kosslyn and his colleagues found activity in area 17 of the brain when subjects were instructed to visualize. Interestingly, area 17 is the first region to receive images detected by the eyes. When people close their eyes and imagine, the brain actually projects the images as if they were seen.*

The Monk's Journey easily yields to the right kind of mental movie. Imagine the monk climbing up the mountain on the first day, starting at dawn and proceeding step by step up the trail. Now imagine the monk climbing down the next day, again starting at dawn and plodding down the trail. Now take the two movies and superimpose them, so that they run at the same time. Gradually the monk coming up and the monk coming down approach one another. At some point, they have to pass one another. At the moment when the second-day monk going down meets the first-day monk going up, the monk is at the same place at the same time on both days. Of course, there is no way of predicting in advance exactly when this moment will occur, but it must occur someplace, because the second-day-down-going monk has to pass the first-day-up-coming monk.

The Four Beetles. Often a good solution to a problem comes by way of guessing and then examining the guess to try to justify it. The Four Beetles includes an important clue, the information that no technical mathematics is needed. This suggests that the answer must be very simple. The simplest answer at hand is the side of the square, one foot. But how can one test and justify such an answer?

It is helpful to substitute a different mental movie for the one given. The aim is to see how the beetles could "use up" the initial distance of 1 foot that stands between them, so never mind the beetles walking. Imagine instead that they stand on ice. A string 1 foot long harnessed to each beetle stretches to the beetle behind, which holds the string in his mouth. At the word "go," each beetle starts to eat the string. As each eats, he is pulled toward the beetle in front of him, sliding instead of walking but with essentially the same path of motion. When the beetles meet in the *center of the square, they have each eaten the whole one-foot string, having therefore traveled 1 foot. (I have discussed this problem and how to solve it at greater length in* Knowledge as Design, *there are a couple of further fine points to consider for a complete argument.)*

The Art and Craft of Reframing

Where do new and better representations come from? The answer is that they themselves must be searched for. Finding a better representation for a problem is a problem in itself. Indeed, it's a problem of planning, a search for a new plan of attack for the original problem. In their *Human Problem Solving*, Alan Newell and Herbert Simon write of "planning spaces" that sit alongside direct approaches to solving a problem. A search of possibilities in a planning space may yield concrete plans or approaches that can then be applied to addressing the original problem.

The catch, of course, is that finding a better representation is itself likely to be unreasonable, and not necessarily any easier than trying to solve the problem as originally seen. The search commonly occurs in a Klondike wilderness. One cannot simply reason one's way to a better representation. One must explore a range of possibilities, traversing plateaus and escaping from canyons and oases along the way. Whether this is any easier than the original problem is hard to predict in advance. But sometimes it is!

While there is no formula for finding a good representation, a look back at the cases just discussed offers some clues about the process. In Giraffes and Ostriches, the problem solver must seek a way to take advantage of the given information, the thirty eyes and forty-four legs. Thinking of constructing fronts of giraffes and ostriches is just one way of accommodating the information. Another might be a process of erasure rather than construction. Suppose one erases thirty eyes and thirty front legs to match the eyes. That leaves fourteen hind legs, which must belong to seven giraffes.

The superimposed mental movies of The Monk's Journey might come from an effort to compare the monk climbing up and the monk climbing down. It's commonplace to juxtapose things in order to compare them—a chip of paint with a paint sample, an authentic dollar bill with one that might be counterfeit, a family photo with another snapped on the same occasion. As to The Four Beetles, imagine a beetle's first step toward the next one. The beetle uses up one step's worth of the side of the square. To be sure, the beetle in front of him moves some, changing the first beetle's direction a little, but this does not seem to be that much of an adjustment. Elaborating this idea leads to the mental movie of each beetle eating a string stretching to the beetle in front of him.

As to Copernicus, his shift from a geocentric to a heliocentric model was fundamental given the philosophy of the times but not that radical in representational terms. Circles were still the elements, everything depending on what they circled around. The case of Einstein might seem the most remarkable. Yet Einstein's idea of traveling along beside a light wave is one instance of a strategy used over and over again in inquiry—getting something to stand still. Strobe lights are used to freeze the motion of dripping faucets or whirring fan blades in the same spirit.

None of this implies that good representations are easily found. Worse, a new representation that seems intuitively promising may take a great while to pursue and then come to nothing. However, the examples all challenge the notion that good representations come utterly out of the air. They typically involve adopting and adapting familiar representational moves, with generous use of analogy. One can be smart about searching for promising ways to reframe a problem.

8

Anything but That

Thinking under Pressure

The story is told of the student of physics who faced this problem on an exam:

✎ The Barometer
Explain how you can use a barometer to measure the height of a tall building.

The student wrote the answer, "Take the barometer to the top of the building, attach a long rope to it, lower the barometer to the street, and then bring it up, measuring the length of the rope. The length of the rope is the height of the building."

This certainly answered the question, but it was not at all what the professor had in mind. The barometer amounted to a hint to use air pressure. At the top of the skyscraper, the air pressure would measure lower than at the bottom. The difference in air pressure would allow one to calculate the height of the building.

The professor puzzled over how to handle the student's

rebellious response. A process of arbitration resulted: Another professor was brought in as judge, and the student was asked to retry the problem, being sure to show knowledge of physics. In the end, the student provided several answers, none involving air pressure, and received full credit. The answers included:

• Take the barometer to the top of the building, and lean over the edge of the roof. Drop the barometer, timing its fall with a stopwatch. Then using the formula $S = \frac{1}{2}at^2$ (which says distance fallen equals one-half the acceleration of gravity times the square of the time elapsed), calculate the height of the building.
• Take the barometer out on a sunny day and measure the height of the barometer, the length of its shadow, and the length of the shadow of the building, and by the use of simple proportion, determine the height of the building.
• Take the barometer and begin to walk up the stairs. As you climb the stairs, you mark off the length [vertically, using the barometer] and this will give the height of the building in barometer units.
• Tie the barometer to the end of a string, swing it as a pendulum, and determine with great accuracy the value of g (the acceleration of gravity) at the street level and at the top of the building. From the difference between the two values of g, the height of the building can, in principle, be calculated.
• Take the barometer to the basement and knock on the superintendent's door. When the superintendent answers, speak to him as follows: "Dear Mr. Superintendent, here I have a very fine barometer. If you will tell me the height of this building, I will give you this barometer . . ."

This tale is a good reminder that not all problems come with one right answer, not even problems that are supposed to have one right answer. Texas A & M investigator of creativity Ronald Finke offers a useful distinction here. He contrasts *convergent insight* with *divergent insight*. Both call for breakthrough thinking, but convergent insight applies when there is one difficult-to-attain answer, divergent insight when there are many.

The trouble is, when you already know one specific answer, or even just the kind of answer you're supposed to have, that tends to shut out others. This is the oasis trap. The answer to it is decentering. Problems are often more open-ended than they look. The possibilities are endless, and the challenge is to find the good needles in the haystack of mediocrity.

Decentering, Fourth Century B.C.E.

The tale of Alexander the Great and the Gordian knot comes down through history as a classic example of decentering from a conventional solution by changing the rules. Greek legend says that the knot in question was tied by Gordius, king of Phrygia. An oracle predicted that anyone who managed to undo the Gordian knot would come to rule all of Asia. Alexander saw his destiny in the prophecy. However, the knot had a formidable reputation for stumping aspirants. Not one to hover over fine points, Alexander wielded his sword and cut the knot right through, "untying" it.

Alexander might have agreed with Einstein, who had this to say about problems:

> The formulation of a problem is often more essential than its solution, which may be merely a matter of mathematical

or experimental skill. To raise new questions, new possibilities, to regard old questions from a new angle, requires creative imagination and marks real advance in science.

Certainly Alexander viewed an old question from a new
angle, although he was less interested in science and more
in conquest. Alexander's tactic both impresses and annoys.
One can admire Alexander's decisiveness and his easy stroll
past the conventional boundaries. At the same time,
Alexander cheated. Whatever the official prophecy said,
even if it only spoke of "undoing," surely slicing the knot
was not what the gods had in mind. Or was it? Perhaps this
was just the sort of no-nonsense initiative that the gods on
top of Mount Olympus sought. In any case, Alexander
went on to conquer Asia and a good deal more.

This tale of Alexander the Great is a paradigmatic example of what psychologists sometimes call *problem finding*.
The silent partner to problem solving, problem finding
gives a name to the idea that problems need to be detected
in the first place and often need redefining along the way
to a solution. Initial visions of the solution, however vague,
can prove to be tempting oases that keep one away from a
better solution or simply an alternative and interesting
solution.

Decentering in the Modern Era

Another pointed tale of problem finding concerns the
early space initiatives of the National Aeronautics and
Space Administration. To recover astronauts safely from
orbit, NASA had to face the vexing problem of reentry.
The heat generated by air friction as a space vehicle

returned from the atmosphere at a velocity of about 5 miles per second would vaporize most materials, including astronauts. Early on, NASA defined the problem as one of finding a substance that would withstand a temperature of 3,500 degrees Fahrenheit, a clear vision of a certain kind of solution. Considerable money was spent trying to find such a substance, but to no avail. Somewhere along the line, somebody decentered to define the problem in a more general way. It wasn't to find a substance with these resistive qualities, but to keep the astronauts cool by whatever means.

The reconceived problem led to a truly elegant solution, the ablative heat shield. This ceramic substance gradually burns away during reentry. As the substance vaporizes, the residue streaming behind the space craft carries the heat away from the craft and the astronauts. Thus the ultimate solution functioned in a way opposite to the initial specification: Far from resisting, it yielded, but in yielding it solved the core problem of keeping the astronauts cool.

Another contemporary classic of problem finding concerns the development of the 3M product Post-its. One of the inventors at 3M had been working on a new glue. Unfortunately, the new glue turned out not to adhere very well. The enterprise appeared to be a failure. However, *repurposing*, as discussed in Chapter 1, is an important kind of problem finding. The inventor repurposed the adhesive, asking what an adhesive that didn't stick very well might be good for. An initial application was a sticky bulletin board. Unfortunately, along with the notes posted on the board, dirt and dust collected, resulting in the failure of this product. Another 3M inventor found inspiration from his church hymnal. He noticed that bookmarks placed in the hymnal invariably fell out. By applying some of the new glue, he believed the bookmarks would stay in place. The result was Post-its, a hugely successful product.

The Importance of Problem Finding

Problem finding is a broad idea with many variations and can help with roving, detecting, reframing, or decentering. Problem finding can range from initial detection of a problem or opportunity, in situations that present an untroubled surface; to seeking alternative definitions of what is problematic about a challenging situation, including repurposing; to redefining a seemingly well-defined problem with an expected sort of solution, in order to open up new ways of thinking about it—decentering:

THE SPECTRUM OF PROBLEM FINDING

←——————————————————————————————→

| Initial detection | Alternative definitions | Redefinition |

Initial detection aside, the great enemy of problem finding is what might be called *solution mindedness.* People have a strong tendency to generate a quick vision of the nature of a solution and set off from there—the glue does not adhere well enough so try to make it stronger. Most people are reluctant to stand back and ask what the problem really is or in what different ways it might be understood.

Solution mindedness certainly does not serve creativity well. One has to decenter from such premature images of a solution. Considerable evidence shows that problem finding makes up a significant part of creativity. One of the best-known studies was conducted by psychologists Jacob Getzels and Mihaly Csikszentmihalyi in the 1970s. They worked with students at a school of art, watching how the students responded to the request to compose a still life, given various objects to arrange. Besides observing the students, they had the students' products rated for creativity by a panel. Their analysis looked at a number of markers for problem finding—rearranging the materials a lot before beginning, changing directions part way along, finding ways to consider deep human themes as part of the still life, not considering the work to be done even when they were done for the moment. The investigators found that these traits correlated highly with the creativity of the still lifes as rated by the panel and also with general ratings of their creativity provided by their instructors. In addition, the ratings of creativity were highly correlated to the students' degree of problem finding as assessed from interviews with the researchers.

The authors conducted a follow-up study a number of years later to see whether the students that had appeared most creative during their initial study stayed that way. Remarkably, they discovered that those who had engaged in more problem finding were more likely to become professional studio artists recognized for their creative output. However, a subsequent follow-up study showed there was a gradual loss in the predictability of success from the original problem-finding data.

Further evidence for an association between creativity and problem finding has been found in other typically creative fields. In a replication of the Getzels and Csikszentmihalyi study, Michael T. Moore of Georgia Southern University used the same procedures and a similar problem situation with student writers. Again, a relationship was found between the level of problem finding and the originality of an artistic product, a written essay. Those students who explored more in the problem formulation phase and were more willing to change their essays in the solution phase were also rated higher in creativity.

In *Surpassing Ourselves*, a 1993 book on the nature of expertise, psychologists Carl Bereiter and Marlene Scardamalia introduce the idea of *promisingness* toward understanding better how people working in a particular field display more or less sensitivity to potential problems worth pursuing. While outsiders to a field have little chance of appraising what is worth attempting and what not, those with experience can develop a nose for promise. Bereiter and Scardamalia suggest that judgments of promisingness involve three aspects: a direct match to the goal, a match to one's capabilities, and pointers to further possibilities. A businessperson or a physicist seeking new worlds to conquer does well to ask, "Does endeavor X align with what I'd like to accomplish? Am I likely to be capable of bringing it off? If I do, will it lead on to other things?"

Redefining Problems to Escape Old Solutions

Consider the classic Nine Dots problem from Chapter 3. The challenge was to draw four straight lines that passed through all nine dots without lifting one's pencil. This insight puzzle traps many problem solvers with the canyon assumption that the lines should stay within the boundaries of the square. Once free of that assumption, it's easy enough to draw the four lines.

But the solution we already know is a trap too, an oasis

trap. Suppose we exclude it—"anything but that!" Can we cover all nine dots by drawing *fewer* than four straight lines? (More is easy.)

Here is one solution. Focus on the middle column of dots. Fold the two sides over so that the left and right columns of dots touch the middle column (see illustration). Now take a pencil and draw one vertical line, simultaneously marking all three vertical columns. Unfolding the paper reveals that all the dots are crossed by drawing just one line.

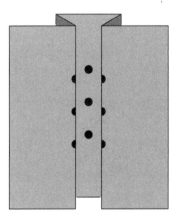

"But you can't fold the paper!" is the natural complaint.

Who says so? Nothing in the problem statement says one can't fold the paper.

"Well, it's understood that you can't fold the paper," is the response.

Is it? Insight puzzles are full of traps where what is tacitly understood is exactly what has to be challenged. Remember The Mask problem from Chapter 2: There's a man at home with a mask. There's a man coming home. What's going on here? You might assume that "home" means a domicile, but that is just an assumption. In fact, it is home base. The rule for insight problems is that everything beyond what is stipulated is open territory.

Here is another barometer solution to The Nine Dots problem. Draw the nine dots at a slight tilt on a piece of paper. Then roll the piece of paper into a cylinder. Flattening the cylinder during the procedure (to keep the line straight in drawing it), draw one continuous straight line that spirals up the cylinder (see illustration). Again, you've drawn one straight line and covered all nine dots.

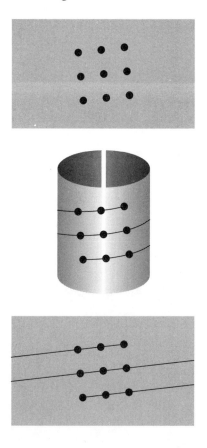

Three Barometer Problems

It's instructive to revisit, in the "barometer" spirit, puzzles from earlier in the book. Here are the first three problems posed in this book, with an added line to exclude the official solution. Are there "barometer" solutions? See what you can find.

✐ Another Coin

Someone brings an old coin to a museum director and offers it for sale. The coin is stamped "540 B.C.E." Instead of considering the purchase, the museum director calls the police, *even though there's nothing wrong with the date.* Why?

✐ Another Sahara

You are driving a jeep through the Sahara desert. You encounter someone lying face down in the sand, dead. There are no tracks anywhere around. There has been no wind for days to destroy tracks. You look in a pack on the person's back. *You do not find an unopened parachute.* What do you find?

✐ Another Mask

There's a man at home with a mask. There's a man coming home. What's going on here?

This is pretty thin information, so perhaps you would like to hear some questions answered. Is the man with the mask a thief? No. Does the man coming home live there? No. Is the man with the mask going to hurt the man coming home? No. *Is it a baseball game?* No.

Now, what's going on here?

As usual, if you want, do your best with these before continuing.

Solutions

What should occur at this point is a revelation of the official answers. However, there are no official answers in barometer country. The best that can be done here is to speculate on some possible answers, and those from readers may be just as good or better.

Another Coin. *The crux of the puzzle is the coin. How could the coin not be a fake despite its 540 B.C.E date? Well, perhaps the coin is a toy coin from an antique game of the 1800s. The museum is a toy museum. Alternatively, perhaps the coin is a rare commemorative coin issued to celebrate something that happened in 540 B.C. The museum is a coin museum.*

That leaves the matter of the director having the seller arrested. Perhaps the director recognizes the coin as stolen, the seller as a criminal. Perhaps the coin is commemorative and properly dated as such, but the seller aims to pass it off as ancient and hence deserves arrest.

Do such answers as these click into place as elegantly as the official answer? Of course not. This is hardly a surprise, since The Coin was written in the first place to fit one answer as well as Cinderella's foot fit the glass slipper. We can't reasonably expect solutions that go beyond the original design of the insight problem to always work as well. The trick is to get them to work at all.

Another Sahara. *What, if not an unopened parachute, might one find in the pack on the man's back? How*

*about a tarpaulin folded to seem like a parachute, mak-
ing the man a murder victim. An investigation should
look into who set him up. Or perhaps the pack contains a whisk broom. Before taking cyanide, the man
has swept the sand clean to conceal his footprints, just
to leave the rest of us with something to puzzle over.
(Of course, he should have buried the whisk broom in
the sand to maintain the mystery longer.)*

*Another Mask. Perhaps it's a softball game. Or perhaps
it's the man's birthday: His brother—and others—are
waiting for him at home in costumes for a surprise party.
Alternatively, maybe the man coming home is a disaster
victim, returning to a flood site. Meanwhile, a diver is at
work seeking to recover valuables from his home.*

Again, do these ideas about Another Sahara and Another
Mask offer the kinds of cognitive snaps provided by the
original official answers? No they don't, but that would be
expecting a lot. The important moral is that the official
answer can be challenged. In the Klondike world, the
reach for insight is in good part a battle to escape from
oases of false promise—in this case, false not so much in
the sense of mistaken as in the sense of used up: "Been
there, done that!" The hardest part of seeking a new
insight can be turning away from a well-worn but satisfactory resolution.

Finding the Real Problem

An especially powerful problem-finding question asks,
What's the real problem? Faced with any particular prob-

lem or any chaotic situation, people can often arrive at answers that decenter from initial expectations about the character of the solution and yield better ones. The question has a paradoxical character, because of course there is no one "real problem." What the problem is can be construed in many ways. This granted, the question nudges the mind in a useful direction—toward what is more fundamental about the situation, toward deeper and more general ways of articulating what the problem might be.

Asking about the real problem is a strategy readily put to work in problem situations. Here are two that readers might like to try. Consider how the real problem might be defined and what solutions that suggests.

✐ The Elevator

This problem is often related by the noted educator and champion of critical and creative thinking Arthur Costa. An office building suffered from a chronically slow elevator. Tenants were forever fretting about the problem, and a bundle of complaints reached the building management, who called in experts. They assessed the building structure and determined that replacing the elevator with a newer, faster model would require deconstructing and reconstructing a good part of the building—not a financially viable option. What is the real problem, and what might be done about it?

✐ The Deer

A certain research institute was located in the midst of a large tract of woodland. A single road with a fence along each side led from the gate to the buildings of the institute. Every now and then, commuters driving along the road would run into a deer. A look at the records showed that this tended to happen around dusk. What is the real problem, and what might be done about it?

Solutions

Although open-ended problems cannot have unequivocally correct answers, certainly possibilities can be offered. Here are some comments on the two problems.

The Elevator. *This episode actually occurred, so there is a specific answer—not the only possible answer but the one that occurred. Casting about for another angle on the problem, the management engaged a different architect. This architect sized up the situation and offered a solution that proved remarkably successful and cost only a few dollars. The management called in carpenters to mount a large mirror in the elevator lobby. People spent time admiring themselves, adjusting their hair, and dusting off their jackets while waiting for the elevator. The operational test of this peculiar solution's adequacy was simple: The complaints to management largely ceased. Like the tale of Alexander the Great, this story also illustrates the power of redefining the problem—from "the elevator is too slow" to something like "people are bored."*

However, it is worth thinking of other problem definitions leading to other solutions. Here are some possibilities for what the real problem is:

- *People use the elevator too much. Make the stairwell especially attractive, and introduce a health campaign to encourage tenants to walk.*

- *The elevator is slowed by people taking it down as*

well as up. Put up humorous "Do your body a favor" signs encouraging people to take the stairs down.

- *People are uncomfortable waiting standing up. Put chairs and benches about the lobby, and they'll wait with more patience.*

- *Perhaps there's a flood of traffic at certain times of day. Suggest that the businesses in the office building stagger their hours a little. It might not inconvenience them much, and it might alleviate the bottleneck.*

- *Perhaps the problem is that the business offices with the most traffic are scattered throughout the building. Consider engaging their cooperation to relocate over several months to offices closer to the street level.*

The Deer. *When people encounter this problem, they tend to fall quickly into a particular view of it and jump from there to a ready solution. For example, someone might take it for granted that the problem is a low fence, so the solution is to make the fence higher. However, a high fence impairs the free movement of the deer across their terrain and ruins the lovely view of the forest as people drive to and from work. Then how else might the real problem be described? Here are some possibilities for what the real problem is:*

- *The deer are more active and less readily avoid the cars at dusk. Adjust commuting hours so that fewer people travel then.*

- *The deer live in a restricted space that's dangerous for them. Pleasant as they are to have on the land, relocate them.*

- *The deer cannot easily see the cars. Post signs at dusk reminding drivers to turn on their headlights.*

- *Drivers drive too fast for the deer to dodge. Post lower speed limits, and install speed bumps.*

- *The collisions might only happen at certain regions. Determine where these are, and install speed bumps.*

Opportunistic Invention

A few years ago, my wife, Ann, asked the family a simple riddle, "What's more beautiful when it's wet than dry?" The official answer was a rainbow: beautiful when wet, and when dry, no rainbow at all. But my oldest son, Ted, quickly provided a very different answer, "Bo Derek!"

Not only is Bo Derek prettier wet than dry, but the bow of the rainbow is even there in the form of Bo. Ted's quip turned a not-very-challenging riddle into an insight problem. In general, spontaneous moments of humor are often opportunistic insights.

Out-of-the-blue discoveries small (like this one) and large (like Archimedes' discovery of the principle of water displacement) are commonplace in invention. However, opportunistic humor has an interesting peculiarity. Ted's mind seemed to have been tuned in advance, not to humor about rainbows, Bo Derek, or anything specific, but to humor in general.

There seems to be a general pattern of alertness at work in such cases, one that could be called the *perverse interpretation pattern.* Anyone who has raised children will know that youngsters often deliberately misconstrue a meaning and convert it to another meaning not intended. Father says, "No TV before homework." The ten-year-old says, "Unfair! It's *always* before homework. There's tomorrow's homework and the day after that and the day after that." Ten-year-olds are quick to take such opportunities. They have their pattern for perverse interpretation primed. Perverse interpretation is a clear case of responding to an oasis trap by decentering from the conventional interpretation.

Children learn the perverse interpretation pattern young, and it sticks and functions as an engine of quick humor in all sorts of situations. The Bo Derek response to the rainbow riddle is a perverse interpretation of the intent of the riddle. I have no idea what went on in Ted's mind, but I suppose it was something like this: "More beautiful when wet? Well, rainbows obviously. Yeah yeah, but what else could it be? Bows . . . Bo, ah! Bo Derek." The perverse interpretation pattern could not guarantee that Ted would come up with something, but it could stimulate him to try.

Most of the cases of sudden discovery discussed earlier have involved people focused on particular problems. But it's not always that way. One can be alert not just for information relevant to particular problems but also for opportunities of a general kind—for humor or for other things. Poets and artists sometimes find the seeds of poems or paintings in the strangest places because their general poetic or artistic mousetraps are set. When this is so, simply getting about in the world amounts to a kind of free search through a Klondike space. It brings one into contact with many occasions that might trigger patterns and become opportunities.

It is understandable that people can acquire rapid reflexes for routine matters—snatching hands away from hot stoves and the like. More provocative is the observation that the right reflexes can promote insightful connection making. Even the animal kingdom offers evidence of just this. Conventional training of porpoises uses standard conditioning techniques. The porpoise gets a reward for doing a stunt well. Initial training involves rewards for approximations of the stunt, later training rewards for refined and reliable executions of it.

But what if dolphins received rewards for inventing new stunts? Karen Pryor, Richard Haag, and Joe O'Reilly conducted just such an experiment at Sea Life Park in Hawaii. They instituted a schedule of rewards for divergent behavior in a rough-toothed dolphin, behavior that might get turned into stunts. Could the dolphin learn such an abstract concept? It turns out that it could. After a while, it caught on and become more divergently playful, cooking up odd moves like aerial flips and swimming in corkscrew patterns that then could be turned into new stunts for the public shows. Even dolphins can solve unreasonable Klondike problems!

Part 3

Mind, Brain, and Breakthrough Thinking

In which we look at the psychological side of breakthrough thinking: whether it's really different from normal problem solving; whether the mind and brain go into some special overdrive during breakthroughs; and how knowledge, memory, and pattern recognition figure in breakthrough thinking.

9

Breakthroughs on Trial

Witnesses for the Prosecution

Archimedes in his bath, Darwin reading Malthus, Gutenberg attending the wine festival—time and again in this book, we have looked at moments when individuals appeared to arrive at fundamental breakthroughs through sudden leaps. Working from these and other cases, the argument has been developed that systematic breakthrough thinking is distinctive. It involves a kind of Klondike logic quite different from normal sequential reasoning.

Yet four cognitive psychologists, including Nobel prize winner Herbert Simon, have advanced a different view. They do not deal with the ideas about breakthrough thinking introduced in this book—indeed, those concepts had not even been articulated when they did their work. However, investigating scientific reasoning in the history of physics and chemistry, they suggest that breakthroughs in science reflect "ordinary problem solving." The results may be extraordinary, but only because of the extraordinary knowledge and intelligence of the scientists responsible for the discoveries. In terms of process, it's business as usual—ordinary problem solving.

Pat Langley, Herbert Simon, Gary Bradshaw, and Jan Zytkow—hereafter called The Four—sum up their perspective in a 1987 book appropriately entitled *Scientific Discovery*. Quite rightly, they warn against the tendency to mythologize the advance of science with tales of sudden discovery enabled by mysterious mental powers. They seek to show that what might look like sudden discovery can generally be explained as a more incremental, progressive kind of problem solving—what this book calls problem solving in a homing space, where the problem solver can home in incrementally on a solution.

The cases The Four examine span an impressive range of kinds of discovery in science. They demonstrate how artificial intelligence programs can work from tables of numerical data to discover important formulas, work from more complex tables of data to propose intrinsic properties not originally present (for instance, mass) and discover formulas based on those, derive general qualitative descriptive rules about phenomena, determine from chemical reaction data which substances are elements and which elements go into various compound substances, and more. Their studies certainly do much to demythologize scientific discovery. However, do the results really show that sequential reasoning rather than breakthrough thinking accounts for scientific discovery? Let us look at three cases in point, three witnesses for the prosecution—Johannes Kepler's discoveries about planetary motion, Max Planck's derivation of a formula for blackbody radiation, and Lavoisier's theory of oxygen—and see how well their testimony holds up.

Kepler as Computer

One of the most impressive products of scientific inquiry is a complex formula that synthesizes the relationships between important variables in a situation. Such a formula seems a huge leap beyond data. Nowhere in a table of data can one read off an algebraic relationship. The formula emerges like the rabbit out of the magician's hat. Perhaps, one might conjecture, breakthrough thinking is at work.

But perhaps not. In their most successful argument for the role of sequential rather than breakthrough thinking in scientific discovery, The Four show how a table of data often allows reasoning one's way systematically to a formula that expresses the key relationships.

For one case in point, The Four examined Kepler's third law of planetary motion. In 1618, Kepler derived his third law, which says that the cube of a planet's distance from the Sun is proportional to the square of its period of rotation around the Sun. In formula form, $D^3 = C \times P^2$ (where C is some constant), or equivalently $D^3/P^2 = C$. Given such a table of numbers, The Four show how one can derive this complex relationship. They do not mean to claim that this is what Kepler did—more that this is a kind of reasoning that can be and has often been done.

The starting point is the first two data columns of the table shown here (initially the last three columns are blank), recording for three planets A, B, and C their distances from the Sun and their periods. For simplicity, A's distance from the Sun and its period are used as units and those of the other planets expressed in terms of them.

A sequence of strategic steps leads from the first two data columns to the hidden formula. The basic goal is to

Planet	Distance, D	Period, P	D/P	D^2/P	D^3/P^2
A	1.0	1.0	1.0	1.0	1.0
B	4.0	8.0	0.5	2.0	1.0
C	9.0	27.0	0.333	3.0	1.0

do arithmetic that combines columns to try to arrive at a constant column, one with all the numbers identical:

1. *When two values increase together, divide them, because a constant might result.* Distance and period increase together from row A to B to C. Therefore, divide them to create the D/P column. As it happens, a constant does not result, but the numbers vary less, and so are closer to a constant.

2. *When one value increases as another decreases, multiply them, because a constant might result.* As distance gets bigger from A to B to C, the new column D/P gets smaller. So multiply D and D/P to get a new column, D^2/P. However, this column is still not constant.

3. *Apply strategy 2 again:* As the new D^2/P column increases, D/P decreases. Therefore, multiply them, getting D^3/P^2. At last, a constant!

4. If a constant column results, this yields a formula in the form: the arithmetic done equals the constant. In this case, the formula reflecting the arithmetic is $D^3/P^2 = 1.0$.

Of course, one could apply these strategies to other combinations of columns, not finding the answer quite so directly. However, there are not many possibilities to search through in order to home in on a constant column.

The Four describe a sophisticated artificial intelligence

program called BACON, named after the seventeenth-century British philosopher who sought to give general rules for scientific reasoning from observation. BACON reasons as illustrated above, using additional strategies to handle more difficult cases. A table of data may seem far removed from a complex formula capturing its regularities, but BACON's success makes the point that systematic patterns of reasoning can bridge the gap from one to another by following numerical clues in the pattern of the numbers—no leap of insight needed.

Max Planck in a Reasonable Mood

On October 7, 1900, the physicist Max Planck derived a crucial formula for a physical phenomenon called blackbody radiation. What blackbody radiation specifically concerns need not worry us here. Suffice to say that the problem Planck solved was widely recognized in 1900, and considerable empirical data had been gathered about it. A formula was needed to predict the intensity of light at any frequency of light. The point of departure was a formula proposed by the German theoretical physicist Wilhelm Wien in 1896.

Planck in 1899 thought he had demonstrated the validity of this formula, reasoning from the laws of classical physics. Unfortunately, accumulating results from empirical researchers showed that the formula was mistaken after all. On October 7, 1900, an experimental physicist communicated to Planck new data on certain extreme cases for which Wien's formula clearly failed. In the range of these new data, another, simpler formula seemed to hold.

By that evening, Planck constructed a new formula compatible both with Wien's formula and the extreme-case formula, interpolating between the two. Planck's formula proved pivotal in the development of physics. Over the coming years, it was recognized that the rationale for the formula would require quantum concepts. The formula contributed to stimulating the quantum revolution. All these circumstances suggest an episode of breakthrough thinking. However, The Four caution against thinking that Planck made some kind of leap. Given the information at Planck's disposal, they suggest that any competent mathematical physicist could reason out the target formula relatively quickly. To prove their point, they presented the two original formulas to eight contemporary physical scientists and applied mathematicians. The Four did not identify the source of the problem. Five of the eight scientists gave the same answer as Planck, and in under two minutes. None of them recognized that this was the blackbody radiation problem before deriving Planck's final formula. Even though Planck's solution represented a breakthrough from a historical perspective, for Planck the problem was a homing problem.

Sometimes it's argued that shifts in problem definition and problem representation—what we called reframing earlier—are key to breakthroughs. The Four acknowledge this, but they emphasize that these sorts of shifts should not be mythologized either. Many creative breakthroughs involve dealing effectively with problems of long standing rather than conceiving the problems in the first place. The puzzle of blackbody radiation was articulated by others decades before Planck solved it. Moreover, seeking a new problem definition or problem representation is itself a kind of problem solving. Success in such endeavors can be counted as one more victory for "normal problem solving."

The Four appear to mount a fundamental challenge to

the place of insight in scientific discovery. However, their message does not say quite what it seems to at first.

Their concerns about sudden discovery laudably target a tendency in the literature on discovery to celebrate moments of mystery rather than offer explanations. Many concepts have been proposed to account for sudden insight that do little more than introduce new nomenclature for insight. The Four have little patience with such obfuscatory maneuvers. Moreover, they effectively demonstrate the homing side of the discovery story: Much more can be done with sequential reasoning than one might suppose at first.

Nonetheless, in my view they lean too far in this direction. Without recognizing a distinction between Klondike and homing spaces, implicitly they suggest that the challenges of science can be met almost entirely through "reasonable" thinking, almost entirely through "normal problem solving."

Such a view goes too far. The Four's focus on homing methods masks genuine Klondike phenomena. The story of Max Planck and blackbody radiation itself offers a good case in point. While Planck's reasoning that day in 1900 may have had a homing character, this was only one step in a larger process. The challenge was not so much finding the right formula as recognizing its significance. Planck discovered the formula by mathematical finesse but still lacked an explanation for why the formula worked. To construct such an explanation required several weeks of work; laboring from October 7 on, Planck had one by December.

The explanation involved the use of probabilities in a way that amounted to the foundations of quantum theory, but Planck did not even recognize this. It was not until around 1905 or 1906 that the significance of Planck's maneuver emerged, pointed out by Albert Einstein and Paul Ehrenfest.

Whereas Planck's discovery of the formula had a strongly homing character, the far slower processes of constructing an explanation and recognizing its significance suggest searches through something closer to a Klondike wilderness.

Lavoisier's Leap

In 1700, fire was thought of as a substance. The predecessor of the oxygen theory of combustion was the phlogiston theory, which emerged during the late 1600s. Phlogiston was part of an explanation for combustion based on the ancient idea of the four elements earth, air, fire, and water. Phlogiston represented fire. The idea was that when something burned, it released phlogiston. One could actually see the phlogiston escaping in the flame. Thus the burning of charcoal could be expressed by a formula something like this:

$$\text{charcoal} + \text{air} \longrightarrow \text{phlogiston} + \text{ash} + \text{air}$$

Scientists of the day thought that the air functioned as an enabling factor, a kind of catalyst. They inferred from reactions like these that carbon was made of ash and phlogiston. Also part of the phlogiston theory was a kind of reaction that occurred in purifying metals from metallic ore, along these lines:

$$\text{iron ore} + \text{charcoal} + \text{air} \longrightarrow \text{iron} + \text{ash} + \text{air}$$

This was interpreted to mean that the charcoal released its phlogiston, which then combined with the iron ore to produce iron. That is, in the course of the reaction, the phlogiston moved over from the charcoal to the iron ore. Iron was therefore made of iron ore and phlogiston.

A curious characteristic of the phlogiston theory was that, from the standpoint of modern chemistry, it had things exactly backward. According to the phlogiston theory, charcoal was a compound of ash and phlogiston, and ash was an elemental substance. But according to modern chemistry, charcoal is an element (carbon), and ash a compound of charcoal and oxygen. According to the phlogiston theory, iron was a compound of iron ore and phlogiston, and iron ore was an elemental substance. But according to modern chemistry, iron is an element, and iron ore a compound of iron and oxygen.

In the late 1700s, the phlogiston theory ran into trouble. Joseph Priestley experimented with reactions in which, in modern terms, mercury oxide released mercury and oxygen. This challenged the idea that metallic oxides were pure elements. Building on this and other ideas, Lavoisier proposed the oxygen theory, putting oxygen in the place it holds today and treating metals as elements rather than compounds.

However, an idea similar to phlogiston lingered on in Lavoisier's explanations. He thought something like phlogiston must exist. (After all, one could see the flame.) It was called "caloric" and held to be a component of oxygen, which was seen as a compound. Thus the transition from phlogiston to caloric—and eventually the dropping of caloric—had a fundamentally revisionary character. What was elemental became compound, what was compound became elemental.

The Four—Langley, Simon, Bradshaw, and Zytkow— present a program called STAHL (named after the German chemist who originated the phlogiston theory) that can reason about chemical reactions like those outlined above, noticing when one substance yields multiple others, when multiple substances combine to form one, and figuring out which substances are elements and which elements make up the various compounds. A flexible program capable of backing out of cul-de-sacs, STAHL can recognize inconsistencies in a particular analysis and seek an alternative analysis.

STAHL is able to demonstrate that phlogiston theory offered a consistent interpretation of a great deal of data available at the time. It is also able to demonstrate that Lavoisier's caloric and oxygen theory offered a coherent interpretation of data. It is even able to show that a hybrid theory involving both phlogiston and caloric and preserving what was taken to be an element and what a compound could account for the data.

However, STAHL is not capable of discovering the need for a shift in theory. The Four comment as follows on the complexity of Lavoisier's reasoning: Lavoisier's data

> were indeed not sufficient for Lavoisier to argue that the caloric . . . came from oxygen gas rather than from charcoal. Lavoisier's belief that oxygen gas contained caloric was based on his earlier idea of caloric as the principle of the gaseous form of matter. He noticed that in many reactions in which the input is heated, some kind of air is disengaged, and that in reactions in which an air is absorbed, fire or heat is disengaged. In short: Fire in, gas out; fire out, gas in. In conclusion: Any gas contains caloric.

And also, "the French chemist made extensive use of generalizations (all acids contain oxygen, all gases contain caloric) and on several occasions accepted the conclusions of a generalization rather than facts that contradicted it."

This account reveals that Lavoisier faced something of a

Klondike situation. Reasoning at the level modeled by STAHL, Lavoisier could have it either way, especially if tolerant of a few discrepancies. STAHL had no reason to shift theories and no mechanism for doing so. Getting to a theory that reversed what was an element and what was a compound involved more sophisticated levels of reasoning that bridged those limits.

Breakthrough Thinking Lives

Johannes Kepler and the orbits of the planets, Max Planck and blackbody radiation, Lavoisier and oxygen are all cases that illuminate the character of scientific discoveries. In the first, according to artificial intelligence simulations, results might have been attained by methodical reasoning from data tables rather than through breakthrough logic (although what Kepler actually did is another matter). In the other two cases, this seems less plausible. More than ordinary reasonable problem solving seems to have been involved.

But none of this should be taken as a rejection of the results offered by Langley, Simon, Bradshaw, and Zytkow. Certainly the work of The Four carries an important lesson. Just because, from an outsider's viewpoint, a discovery emerges relatively quickly, is no reason to presume breakthrough thinking must have been at work. The discoverer may have been following clues through a largely homing possibility space every step of the way. The outsider simply lacks the discoverer's nuanced perception of the terrain, like Watson observing Sherlock Holmes at work. There are more clues that point direction in a table of data or a pair of approximate formulas than the outsider might suppose. Indeed, were there not quite a few homing opportunities, science would stay lost in Klondike wildernesses most of the time.

Nonetheless, complex and realistic situations of discovery involve problem spaces that mix Klondike and homing characteristics. The emphasis in the analyses offered in *Scientific Discovery* falls on the homing end of the spectrum, masking unreasonable Klondike phenomena. But on a close look, it appears that the cognitive snap is alive and well in the world of science, as in other worlds.

10

Is There a Mental Overdrive?

The Mysterious Magical Cognitive Snap

Here's Archimedes in his bath again. Once more we see him out for some relaxation, maybe not so eager to crack his head on Hiero's puzzle or anyone else's. But he can't help it. His mind is saturated with the problem. And the very water speaks to him, spilling over the side as he settles into the basin. Water. Volume. Hiero's crown. It all comes together in a moment of insight, a cognitive snap yielding the principle of displacement.

What's going on here? The Klondike perspective gives a kind of structural account of it—a matter of search through a Klondike wilderness of possibilities with a sudden brief homing episode at the end, in this case triggered by the cue of the overflowing water. But what's going on psychologically, in the mind and the brain, to pull a solution out of a bathtub?

The experience of sudden insight is so extraordinary that it's natural to turn to extraordinary explanations, including inspiration. The term *inspiration* comes from the Latin *inspirare*, "to breathe into." Inspiration means first and foremost an influx of insight breathed into a human being from the gods. The idea of inspiration echoes the common experience that insights arrive out of nowhere, bolts from the blue. Inspiration is the ultimate mental overdrive. With the gradual growth of a science of mind, it became natural to seek the sources of inspiration in complex mental mechanisms rather than in divine sources. Attention shifted to puzzling out how the mind itself rather than a god behind the scenes was the factory and fabricator of insight. The normal processes of reasoning and comprehension seemed inadequate to the sparkling achievement of insight, so something more was sought, some special process, some distinctively psychological overdrive.

One such was Arthur Koestler's notion of bisociation, advanced in his 1964 book *The Act of Creation*. Another, more contemporary example is selective encoding, selective combination, and selective comparison, three cognitive processes foregrounded by Janet Davidson and Robert Sternberg to explain insight. Still another overdrive with a different twist is incubation, where during time away from a problem the mind somehow keeps working quietly like a toaster, eventually to pop up a solution. All three of these concepts will be discussed in greater detail later in the chapter.

Such proposals are tempting. They attach distinctive processes to what is certainly a distinctive kind of human experience. Still, this move has a risk. There is even a heavy-duty word for the risk: *hypostatization*. This means ascribing material existence to something, the catch being that the "something" may not really exist. Our experience of heat and cold provides the perfect example of hasty hypostatization. Early scientists and today's youngsters both lean toward the very natural conclusion that cold is some kind of substance that inheres in objects. But physics teaches that cold is simply the relative absence of heat. Although we experience it as something distinct, it is not.

To turn to a side of human experience closer to insight, consider humor. An earlier chapter explored the analogy between jokes and insight problems. Like insight problems, good jokes deliver a sudden frisson of understanding. To bring those ideas back to life, here is another quip from Boston comic Stephen Wright:

> The planes from the local airport really fly low over my house. I was walking from the kitchen to the living room the other day, and the stewardess told me to sit down.

If this seems funny, it is because the human mind, encountering the joke, can quickly integrate prior knowledge about planes taking off, stewardess instructions, and walking about one's own house to comprehend the anomaly in the situation.

Humor certainly is a distinct human experience. But would we want to posit a special psychological "humor process" to explain how a joke is understood? This seems to go too far. A joke is understood the same way as any story, only once we understand it, we see the contradiction or sudden switch at its heart and laugh. A distinctive experience like humor does not necessarily imply a distinctive humor process triggering the experience.

Likewise, it's worth being cautious about proposing special psychological "insight processes" like bisociation or selective processing or incubation to explain how people suddenly arrive at breakthrough understandings after a long search. Maybe we need to recognize such processes.

But maybe we do not. We don't want to fall into the trap of hasty hypostatizations of supposing there's a special psychological process to match experiences that feel distinctive but may not be. Let us look at the evidence.

Bisociation Good and Bad

In his 1964 book *The Act of Creation*, Arthur Koestler saw ordinary thinking as a train running on familiar tracks. Normal thought operates within a frame of reference, a familiar and established domain like Newtonian physics or Keynesian economics. Problems arise and get solved, opportunities emerge and get taken, but without any leap out of the frame of reference. Koestler's view resembles Thomas Kuhn's theory of scientific revolutions. Kuhn argued that "normal science" proceeded comfortably within what he called a paradigm, extending, elaborating, and enriching the paradigm but not transforming it.

Fundamental invention, Koestler proposed, involves jumping the tracks of the prevailing frame of reference, or paradigm. Indeed, invention characteristically brings into play two frames of reference, combining them in some fashion. To turn back to familiar examples, Archimedes connected the context of water and baths with the context of measurement of volume. Darwin connected the context of evolution with the context of Malthusian spiraling populations. Gutenberg connected the context of printing with the context of wine making.

Koestler's name for this happy joining of separate frames of reference was bisociation. Koestler's notion of bisociation as an accomplishment of insight makes good sense. Often, if not always, acts of insight involve linking together frames of reference that otherwise stand apart. Moreover, the idea of bisociation even makes good Klondike sense. A problem solver in a Klondike space searches through its wilderness in quest of solutions. The problem solver's path often follows the contours of the familiar and convenient—a narrow canyon of exploration that in Koestler's terms is a frame of reference. Only when the problem solver manages to find a way out, connecting the original domain of the problem with a new realm altogether, is the problem solved.

Bringing frames of reference together certainly is an important action. But does it involve a distinctive psychological process? Koestler leans toward that conclusion. He sees ordinary consciousness as guiding ordinary thinking, the stimuli that surround us shaping and channeling our perceptions and actions. In contrast, he suggests that normally restrained subconscious processes accomplish bisociation.

Remember the analogy with humor. There is no need to invent a special humor process to explain how people understand jokes well enough to laugh at them. Likewise, there is no need to posit a special unconscious process to explain how connections are discovered between frames of reference. The idea of bisociation as a kind of connection is sound, and we can thank Koestler for that. The idea of bisociation as a variety of mental overdrive that mostly lies sleeping but sometimes wakes into action is, well, unevidenced, unparsimonious, and unnecessary.

A Selective Process Theory of Insight

Another well-known overdrive theory of insight emerged from a collaboration between Janet Davidson, now of Lewis and Clark College, and Robert Sternberg of Yale

University. The theory also forms part of Sternberg's well-known *triarchic theory of intelligence*, but there is no need to look at the whole of his theory in order to understand their view.

Davidson and Sternberg propose three distinct processes as the mechanisms of insight—selective encoding, selective combination, and selective comparison. *Selective encoding* happens when a thinker sorts through the given information and focuses only on the information relevant to solving the problem. With the help of selective encoding, information that originally seemed irrelevant may come to appear crucial and vice versa. A famous example of selective encoding in science was Ignaz Semmelweis's discovery of the importance of asepsis. While serving as a staff member at a hospital in Vienna, Semmelweis noticed that the women assigned to the poor ward had a higher incidence of death due to infection than the women assigned to the rich ward. He discovered that doctors washed their hands less frequently in the poor ward, which in turn contributed to the spread of puerperal fever.

Selective comparison occurs when the thinker discovers a nonobvious relationship between new information just put on the table and old information already in memory. Here the thinker may employ analogies, metaphors, and models to solve problems. Archimedes' discovery in the public baths is an example. Archimedes connected the in-the-moment experience of settling into the bath with the already-in-memory puzzle of measuring the volume of the king's crown.

Selective combination occurs when the thinker discovers nonobvious pieces of information and combines them to form novel and relevant wholes. Davidson offers Darwin's formulation of the theory of evolution as an example of a selective combination insight. Darwin already had the relevant pieces of information to form his theory, but it was only when he discovered how to put them all together that a coherent theory emerged. Of course, one might also see Darwin's discovery as a case of selective comparison, since Darwin formulated it in response to Malthus's essay on population, where he discussed population pressures and the deaths that would inevitably result.

In Davidson and Sternberg's three-process theory, it's not just *that* selective encoding, comparison, or combination occur, but *how*. One might arrive at selective encodings, comparisons, or combinations by a process of homing in through reasoning. However, this does not count. According to the three-process theory, genuine insight occurs only when a person does not have a routine procedure to solve the problem at hand. The selective encoding, comparison, or combination must not happen immediately upon presentation of the problem, but after a delay and then suddenly, in a manner that changes the thinker's mental representation of the problem.

Davidson illustrates the three processes with three insight puzzles. As usual, readers may want to try them out using systematic breakthrough thinking.

✎ The Checker Players (Selective Encoding)
Two men play five checker games, and each wins an even number of games, with no ties. How is that possible?

✎ Three Steaks (Selective Combination)
George wants to cook three steaks as quickly as possible. Unfortunately, his grill holds only two steaks, and

each steak takes two minutes per side to cook. What is the shortest amount of time in which George can cook his three steaks?

✐ The Garment Rack (Selective Comparison)
Using two boards 3 to 4 feet long and a C-clamp, make a garment rack strong enough to hold a hat and coat. (This problem was used in the classic work on insight by Norman Maier during the 1940s.)

Solutions

Here is how these problems can be seen as examples of selective encoding, combination, and comparison, along with the solutions.

The Checker Players. *The problem solver tends to assume that the players are playing against one another, a faulty encoding suggested by but not stipulated by the problem statement. Once the problem solver re-codes the problem to recognize that the players need not be confronting one another, each can easily win an even number of games playing against other players.*

Three Steaks. *In contrast with The Checker Players, there is nothing misleading about the problem statement here. It's just a matter of combining the given pieces of the situation to solve the problem. The obvious combination has George cooking two steaks on one side (two minutes), then cooking the two on the other side (two more minutes), and then cooking the third steak alone on the grill, first one side and then the other (4 more minutes), for a grand total of 8 minutes. Worth noticing is the fact that George underuti-*

lizes the grill during the last 4 minutes, with only one steak cooking. This suggests searching for another combination that maximizes grill use: two steaks on one side (2 minutes); then set aside one steak, finish the other, and start a raw steak (2 more minutes); then turn the new steak over and bring back the set-aside steak to cook its other side (2 more minutes), for a grand total of 6 minutes and no grill downtime.

The Garment Rack. The classic solution to this problem is to use the C-clamp to fasten the two boards together end to end, with an overlap. This construction can be wedged floor to ceiling. The hat and coat can be hung on the stem of the C-clamp itself.

Davidson notes that this problem might be solved in different ways. Arguably the solution amounts to selective combination, since the problem solver has to combine the materials in the right way. Selective encoding might come into play too, since the problem solver has to encode the C-clamp not only as a fastener but as a potential hook. However, one way of arriving at the solution clearly involves selective comparison. Sifting their minds for models, problem solvers may remember pole lamps and construct the floor-to-ceiling garment rack by analogy.

What Selective Process Proves
and What It Doesn't

Like bisociation, the idea of selective encoding, combination, and comparison is illuminating. The right encoding is one way out of a canyon, the right combination another, and the right comparison with a provocatively different case still another. Accordingly, Davidson and Sternberg's basic concept makes a worthwhile contribution to our understanding of creativity and also sits comfortably with Klondike theory. However, it can be questioned whether selective encoding, combination, and comparison correspond to distinctive cognitive mechanisms—overdrives that kick in now and again. Indeed, as Davidson herself acknowledges, "The three-process theory . . . is mainly a descriptive theory of insight; little is known about what subjects are actually doing or experiencing when they solve the problems." Davidson also notes that it's not known what provokes the sudden use of the processes in nonobvious situations. What the supposed processes do is clear, but when they swing into action and how they operate is not.

Davidson and Sternberg offer various experiments in support of the three-process theory. Several findings are worth mentioning:

- Some people have difficulty knowing when to apply the three insight processes, while others do not. Highly intelligent people were more likely than less intelligent people to spontaneously select and apply relevant information. People of average intelligence do better when clues are given that help the selection.

- The ability to apply the insight processes is fairly highly correlated with IQ.

- People with high IQs are slower, not faster, than people with lower IQs in analyzing the problems and achieving the insights. This shows that insightful problem solving is not always faster problem solving.

• People with higher IQs are more likely to have sudden realizations about an insight problem's solution.

• Insight can be trained on the basis of the three processes, and the training effects are transferable and durable.

It's important to recognize what these findings establish and what they do not. They show that selective encoding, combination, and comparison are helpful concepts for solving insight problems and organizing instruction. However, the findings do not give any evidence for the existence of three cognitive mechanisms contributing individually to insight. In calling it "mainly a descriptive theory," Davidson acknowledges point-blank that selective encoding, combination, and comparison name events that happen during insight, without saying how they happen. As to the training effects, one might interpret Davidson and Sternberg's students as learning to navigate better in Klondike space rather than strengthening some mental overdrive. The ideas of encoding, combination, and comparison tell them something about the structural traps in Klondike space and how to approach such traps.

Incubation: The Pause That Refreshes

For many years, Coca Cola used the slogan "The pause that refreshes." Those days are gone now, but psychologist Robert Olton borrowed the phrase to label one of the most mysterious elements of insight: incubation. The idea begins with the observation that many breakthroughs are preceded by a seemingly inactive period. Recall again the case of Archimedes, puzzled by how to measure the volume of the king's crown but going off to take a bath, or that of Gutenberg, escaping from the dilemmas of printing the Bible to enjoy a wine festival. It's often been suggested that something crucial happens during this time away from the problem when the thinker invests no obvious effort in pursuing it. Whatever occurs during this time to promote arriving at a solution later is called *incubation*.

The simplest question to ask about incubation is whether it truly happens. Does time away from an insight problem help? Interestingly, looking for incubation has proved to be as much of a chore as looking for that tiniest of subatomic particles, the neutrino. Psychologists' efforts to capture incubation in the laboratory generally have failed. The usual experiment is straightforward enough. The experimenter presents an insight puzzle to a number of subjects. Some solve it, some do not. Those subjects who succeed are not studied further; one cannot study incubation with people who have already arrived at an answer. Those subjects who did not solve the problem are divided into two groups. One group gets some time away from the puzzle, along with a request not to work on it during the break. Those in the other group continue to work on the puzzle. The experimenter evaluates which group—the one with the break or the one without—finds the most solutions.

Again and again, the answer has been that it does not matter. The break does not help. In a few studies, a break led to more solutions. However, these sporadic findings have proved hard to replicate.

One might conclude that incubation does not occur at all. However, this would be hasty. There are many reasons why the kind of straightforward experiment described here might not capture incubation at work. To give the phenomenon a better chance, a further question has to be asked. If incubation does occur, how might it operate? There are several different ways a pause might refresh:

Energy recovery. Laboring over a problem generates fatigue. After a break, the problem solver returns with more mental energy.

Musing. While seemingly away from the problem—mowing the lawn, strolling down the street, commuting by bus—the problem solver occasionally and idly pokes at the problem. This casual exploration helps solve the problem not only because of the time investment but also because of its loose, playful style. (If the problem solver fully engages the problem while doing another, undemanding activity like mowing the lawn, this is not incubation in any reasonable sense. It's just time-sharing.)

Fertile forgetting. Time away from the problem lets the problem solver forget about assumptions and approaches that are canyon and oasis traps. These fade over time, and the problem solver returns to the problem with a liberated mind.

Pattern priming. Intense concentration on the problem has sensitized the problem solver to any sort of pattern that might constitute a clue—setting a mental mousetrap, so to speak. Doing other things brings the problem solver into contact with clues that never would have been encountered during focused work on the problem.

Unconscious mental marathons. The problem solver's unconscious mind does serious extended thinking on a problem with consciousness directed elsewhere. In contrast with musing, which involves brief conscious episodes while doing something else, this explanation proposes extended unconscious episodes of focused problem solving while doing something else.

It's surprising to note how many ways incubation might occur. If real at all, incubation could reflect not one mechanism but several. The mechanisms also vary in whether they would be detected by the kind of experiment outlined earlier. Energy recovery, fertile forgetting, or unconscious mental marathons might well appear, but only if the break were long enough. As to musing, subjects would normally be told not to think about the problem during the break. Pattern priming would help only if the break included encouragement and opportunity to explore environments that happened to contain clues. This easily might not be the case. Therefore, one should not conclude on the basis of negative experimental results that incubation does not occur. The experiments might inadvertently exclude the very mechanisms by which incubation usually happens.

Even without experiments, some conclusions are reasonable based on common experience. Energy recovery and musing surely are real factors from time to time. As to the first, cycles of fatigue and refreshment are plainly part of the human condition. As to musing, it is a kind of work on the problem. Maybe the free style of musing has added value or maybe the musing just provides more time-on-task, but either way there should be a benefit. Fertile forgetting sounds like a plausible mechanism, and direct evidence for fertile forgetting will be offered in Chapter 12. Pattern priming sounds like just what happened to Archimedes, Gutenberg, and many others who have fortuitously encountered environmental cues that sparked insights. Recall that Archimedes noted the water spilling over the sides of the tub, and Gutenberg saw a wine press at the wine festival.

An unconscious mental marathon is a different matter. This surely falls into the category of mental overdrives. It's not what we take to be part of the normal functioning of the mind. Yet very often, this is what people have in mind when they speak of incubation. Is there any real reason to acknowledge such a mechanism?

The idea of unconscious mental marathons is immensely attractive. While a person struggling with a problem goes about the ordinary business of life, or perhaps while the person sleeps, the mind continues surreptitiously to compute hour after hour, doing the work that might have been done deliberately and maybe doing it better. Many people feel that something like this has happened when they wake up with a good idea or come back to a problem and seem to find a solution waiting for them.

The problem with attributing such successes to unconscious mental marathons is that there are many other reasons why time away from a problem might help, as already discussed. However, unconscious mental marathons do offer something special. They propose a mechanism by which a great deal of reasoning might get done, something none of the other theories of incubation offer. Therefore, it's essential to ask a very basic question: Do typical insights require a great deal of reasoning?

Actually, no. The central argument of this book is that breakthrough insights do not typically require a great deal of reasoning. According to the Klondike view of insight, solutions are hard to get to, lost in a wilderness of possibilities amid clueless plateaus, hidden beyond entrapping canyons and tempting oases. However, once the problem solver is close, he or she encounters a small homing space inside the Klondike space that leads quickly to the solution, often in a moment of comprehension or at worst with a few steps of reasoning. Sudden out-of-the-blue discoveries appear to be due not to extended unconscious reasoning but to prior saturation in a Klondike problem, priming the mind to detect a lucky clue when it shows up.

In summary, the trouble with the idea of the *unconscious mental marathon* is simply that it does not seem necessary to explain anything. There is nothing for it to do. Not complex reasoning, because typical cognitive snaps do not need complex reasoning. Not general casting about either, because that gets done in other ways.

Keeping It Simple

Throughout this chapter, I've urged caution about positing special-purpose psychological or brain mechanisms to account for the cognitive snap of sudden insight. At the structural level, the snap is simply a consequence of problem solving in the wilderness of a Klondike possibility space—a wilderness of many possibilities, clueless plateaus, narrow canyons of exploration, and oases of false promise. The snap is simply a rapid episode of homing in on a solution, once one gets close enough to a solution so that there *are* clues to lead the way. At the psychological level, we do not need to posit special-purpose insight processes like bisociation or selective processing or unconscious mental marathons to account for how the mind achieves this sudden closure, any more than we need to posit a special humor process to explain how the mind understands jokes well enough to see their twist and laugh at them.

A principle of parsimony rules here. Articulated by the

fourteenth-century monk, theologian, and philosopher William of Ockham, this principle is widely known as *Ockham's razor.* Ockham basically said, "When explaining something, keep it as simple as you can. Don't assume anything more than you need to." Ockham's razor applied to the cognitive snap would say, "Don't assume the existence of mental overdrives if you don't need them to explain the observed phenomena."

So what mental mechanisms do account for cognitive snaps? Two familiar and entirely nontechnical psychological resources—reasoning and comprehension—help the problem solver to take those final steps. Ordinary reasoning explains what was called slow snaps. A problem solver faces an insight problem, finally discovers a reasonable approach, and works out and confirms the value of a solution over the course of a minute or two, or an hour or two—fairly quickly compared with the time spent searching through a wilderness of possibilities.

As for quick snaps, the true frissons, ordinary comprehension provides the psychological mechanism. *Comprehension* is an everyday name for the mind's capacity to make quick sense of something, given supportive stimuli and the right background knowledge. You hear the comment, you comprehend it. You hear the joke, you comprehend it. Likewise, when you're attuned to a problem, poking around in its complexities, familiar with its ins and outs, attentive to clues, open to unexpected directions in true Klondike spirit, you come across the right internal or external cue toward a solution . . . and suddenly you comprehend it.

This is not to say that comprehension itself is well understood. On the contrary, it's quite a mystery. Psychologists, philosophers, researchers in artificial intelligence, and many others have puzzled over the nature of comprehension.

However, the agenda of the moment is not to explain comprehension but to identify psychological resources involved in sudden insight. The argument is that ordinary comprehension is one of them. Comprehension seems to work no differently in achieving sudden insight than in understanding more ordinary matters like a comment or a joke. It puts the picture together for us quickly and intuitively. In the case of insight, the result surprises us so much we are tempted to posit a special process—hypostatization again. But we should not. Ockham's razor again.

What's truly extraordinary about insight comes from the Klondike topography with its surprise at the end, not any exotic mental overdrive. The art of breakthrough thinking, explored in Part II, does not involve activating a mental overdrive but simply problem solving in ways strategically adapted to Klondike possibility spaces.

11

Presence of Mind

Wise Eyes

Sherlock Holmes, Sir Arthur Conan Doyle's famous detective, had a remarkably agile and logical mind. But on top of that, he cultivated a voluminous knowledge of cigarette ashes and the kinds of soil to be found around and about London. Holmes's dedication to knowledge was very selective. On one occasion, Watson discovered that Holmes had an appalling gap in his astronomical knowledge. He did not know that the Moon circled the Earth. Holmes dismissed this gaffe. Such information simply was not important knowledge for him.

Holmes's attitude toward knowledge honors a simple fact: Problems look different when one knows a lot about them. The scene of a crime is an open book to a detective knowledgeable about soils, cigarette ashes, and other appropriate arcana. Imagine Watson inspecting the same crime scene as Holmes. To be sure, Watson was a less acute thinker, but he was also far less informed about what to look for and what it would mean. Recall that detecting subtle cues is one of the hallmarks of breakthrough thinking, a response to seemingly clueless plateaus. Because of what knew, Holmes was in a position to read the scene of the crime for clues that cracked the case. Watson was always stuck on clueless plateaus.

To generalize, knowledge sharpens the eye of the beholder. Lack of the right knowledge produces blindspots for breakthrough thinking. Having the right knowledge and having it handy is part of what prepares one for insight. The last chapter made the argument that we should not favor the exotic in our quest for a psychological understanding of breakthrough thinking. There was no need to posit mental overdrives that roared into action specifically in the service of insight. On the contrary, familiar mental mechanisms like reasoning and comprehending achieve breakthroughs under the right conditions.

But more can certainly be said about the nature of those conditions. What makes knowledge ready at hand rather than stored away in some mental attic and rarely dusted off? How do we best understand the kind of mental readiness that appears to prepare one for sudden recognition of significant clues? Is there any evidence that such mental readiness is real? Questions such as these ask about a presence of mind that prepares us for moments of breakthrough.

The Challenge of Inert Knowledge

Knowing what you need to know is inevitably important. Unfortunately, knowing is not enough. Besides having the right knowledge, one has to retrieve it. Problem solvers often know enough but fail to access what they know. Many insight puzzles are crafted to encourage such failures of retrieval, including the following example.

✎ The Polygamist

A fellow made his home in a little town. In the course of time, he married ten women who also lived in this town. He never divorced any of them, nor broke any law, nor did any of them die. How could this be?

Puzzles like this bias the problem solver to keep the relevant knowledge buried in storage. They create a canyon of misinterpretation from the first. Recall The Mask, which said that there was a man with a mask at home and a man coming home and asked what was going on. The puzzle statement evokes images of domiciles and criminals, blocking retrieval of the real scenario, a baseball game.

As to The Polygamist, one might think that he made his home in another country with very liberal laws or belonged to a distinctive religious group. But no, the intended solution is that the man was a minister. He married each of the ten women to different husbands. Like The Mask, The Polygamist depends on misdirection. The word, *marry* has two closely related meanings, "to get married" and "to perform the ceremony." The wording of the puzzle evokes the first and suppresses the second. Everyone knows both meanings perfectly well, but bringing the second on stage is hard when the first already has the spotlight.

Some people solve this puzzle readily, but some do not. Psychologist John Bransford and his colleagues conducted experiments using The Polygamist (phrased slightly differently) and other similar problems to demonstrate how hard it can be to activate *inert knowledge* in the service of problem solving. Their experiments showed that even when a previous simple activity led participants to focus briefly on the fact that ministers married people, still the participants often did not solve the puzzle. Their studies also showed that active use of knowledge for problem solving, in contrast with just having the knowledge in memory, helped to keep the knowledge active and fostered problem solving.

Background Knowledge and Foreground Knowledge

Canyon traps that deliberately (as in puzzle problems) or accidentally (as in real-world problems) induce narrow assumptions and thus mask relevant knowledge are one of many hazards of the Klondike world. But masking aside, it's an inevitable feature of our mental landscapes that some knowledge lies in the foreground—more readily triggered and retrieved in the brain—and some in the background. Knowing is not an absolute, but a matter of degree.

The right foreground knowledge was surely important to such well-known breakthroughs as Darwin's theory of natural selection and Gutenberg's invention of the printing press. In such stories of discovery, it is not difficult to follow the events in retrospect. The basic insights do not depend on any particularly technical expertise. However, were you or I in their situations—Darwin's reading of Malthus, Gutenberg's presence at the wine festival—and even had we been toying with the problems that saturated their lives, we would not be likely to make the connections that they did.

One's foreground knowledge can put one in a more prepared position to attain an insight. A recent experience with an insight puzzle provides a wonderful illustration. I posed three or four insight puzzles to a friend I'll call Carl.

He solved them all easily. Then I offered The Fan. Recall that in this story, a man fell asleep in church and dreamed he was involved in the Boxer Rebellion, captured, and sentenced to beheading. As the axe descended in his dream, his wife noticed him dozing and tapped him briskly on the back of the neck with her fan. The shock, so the story goes, killed him instantly. This story was told to me as true, but it could not be. Why not?

Despite his earlier successes, Carl came upon hard times with The Fan. After he struggled for a while, I offered some canyon advice: Watch out for mistaken assumptions. This helped not at all. Then I offered another piece of canyon advice: Try to take a larger view of the matter, see things in an expanded context. This counsel stimulated a breakthrough. Carl said that originally he had viewed the story from the man's standpoint, the inner narrative of the dream about the Boxer Rebellion. With my clue in mind, he tried experiencing the story from a different viewpoint, the wife's, and immediately realized she had no way of knowing what her husband was dreaming—nor could anyone else.

It was heartening to find that the Klondike counsel helped. However, having habits of mind that suit the problem at hand can help even more. A couple of weeks later, Carl told me he had given the problem to his father, who was a lawyer. Carl's father saw through the problem instantly. He wondered whether the wife could be tried for manslaughter. Had she been negligent? who would the witnesses be?—key questions for any lawyer building a case. Immediately it became obvious that neither the wife nor anyone else could have known what the man was dreaming. The father brought to The Fan a repertoire of foreground knowledge that easily unlocked its secret.

What Alexander Fleming Already Knew

In the fall of 1928, British bacteriologist Alexander Fleming noted an odd pattern of growth in a culture of staphylococcus bacteria in his laboratory. Molds are a pernicious enemy of research into bacterial cultures. They can take hold in the nutrients on culture dishes and spoil the cultures, requiring the researcher to start over again. Just that had happened in this case, and in a provocative way. Where the mold grew, the staphylococcus did not. Indeed, the bacteria seemed to shun the regions taken over by the mold, leaving an infection-free ring around it.

Fleming might well have tossed the culture out and started over. Instead he read the significance of the pattern. It seemed that some substance secreted by the mold was attacking and killing the staphylococcus. This antibiotic agent turned out to have another fortunate property besides killing bacteria: It left normal tissues unharmed. The mold was a variety of *Penicillium*, and the substance later purified from it came to be called penicillin.

Fleming's discovery seems like common sense, a breakthrough accomplished by being at the right place at the right time. Anyone might have made the discovery, at least anyone who dealt with bacterial cultures. However, this is not quite the case. The bare-bones version of the tale obscures how prepared for this discovery Fleming was, how much in the foreground of his mind was the possibility of antibiotic agents.

Fleming had become interested in antibiotics during World War I, where problems of sepsis threatened the lives of wounded servicemen. After the war, he continued his interest with an active program of research. In 1921, he discovered an antibiotic agent found in tears and egg whites, which he called lysozyme. Although lysozyme attacked some bacteria, it unfortunately had negligible impact on bacteria that threatened humans. Still, encouraged in his quest, Fleming continued his inquiries.

When the fateful moment came in 1928, Fleming was not just any one of many researchers dealing with bacterial cultures. He was an individual specifically on the lookout for antibacterial agents produced by other life-forms. The incident of the mold was indeed an accident. However, it was not just a matter of being in the right place at the right time. Fleming was the right person in the right place at the right time.

How Pattern Priming Works

Louis Pasteur, the nineteenth-century French chemist and biologist, inventor of vaccination and pasteurization, made one of the most famous comments in the history of reflections about insight: "In the fields of observation, chance favors only the mind that is prepared." Pasteur's remark gives testimony to the plausibility of being ready to detect a fortuitous clue, what the last chapter called pattern priming.

Stories aside, can pattern priming be brought into the laboratory to verify its reality? The story here begins with the German psychologist Bluma Zeigarnik, who in 1927 identified what has become known as the *Zeigarnik effect*. It's a common experience that an unsolved problem seems to linger on the fringes of one's consciousness. Zeigarnik performed an experiment showing that people's memory for unsolved problems is superior to their memory for solved problems. Something about an unsolved problem stakes out mental space and occupies it, like a stubborn squatter. It will not get up and leave. It demands attention. The Zeigarnik effect aligns nicely with two mechanisms of incubation. One is musing: If an unsolved problem lingers on a problem solver's mind, the problem solver may work on it from time to time in odd moments. The other is pattern priming. With a problem in mind, if the problem solver encounters anything during other activities that constitutes a clue, the problem solver is more likely to recognize that clue and capitalize upon it.

If the Zeigarnik effect proved reliable, this would give the story a quick and happy ending. Unfortunately, further studies by various investigators turned up results akin to those for incubation. The Zeigarnik effect sometimes made an appearance but often did not. Why the variability? Colleen Seifert and colleagues, writing on the nature of insight, suggest that the Zeigarnik effect occurs only when problem solvers are genuinely stuck on a problem. Sometimes they simply give up instead. Also, sometimes problem solvers have not given up but are still in motion, with no stone wall to butt against.

Psychology's technical term for the state of being genuinely stuck is *impasse*. Some contemporary views of learning argue that impasses have a special importance in cognition. They create learning hot spots that stay with the learner until the learner finds a way past them. Think of the avid basketball player or musician who recognizes a technical weak spot and practices compulsively to master it. Much of people's learning of very ordinary skills, such as driving, spelling, or arithmetic, gets organized around the impasses of the moment. This view suggests that problems not just pursued but pursued to an impasse might be more likely to leave mental traces that would be activated by clues encountered later. In the language of Seifert and her colleagues, impasses may create "failure indices," codings in the mind that stand ready to connect later circumstances with the unsolved problem. Certainly such failure indices would be sensible mechanism for effective learning and adaptation.

Is Pattern Priming Effective?

Seifert and her co-researchers set out to test the idea of failure indices directly. They did not use insight problems but something simpler that could still demonstrate the point: straightforward general information questions such as "What is a nautical instrument used in measuring angular distances, especially the altitude of the Sun, Moon, and stars at sea?" The answer of course is "a sextant." Another was "What do you call one of the thousands of small planets between Mars and Jupiter with diameters from a fraction of a mile to nearly 500 miles?" The answer is "an asteroid." The investigators chose general-information questions that called for information the subjects almost certainly knew, but did not use very often. Accordingly, the subjects often would not be able to retrieve the information.

In the first step of the study, the investigators asked a number of subjects these questions and recorded their initial successes and failures. Then about half an hour later, the subjects undertook another task that appeared unrelated to the first. They saw a range of letter strings on a screen and had to decide which letter strings were words (such as umbrella and sextant) and which were not (for example, dascribe and trinsfer). Among the words were some, like sextant, that were answers to the questions the subjects had tried to answer. After this, the subjects took the rest of the day off. Returning the next day, they tried to answer further general-information questions, some of them new and some repeats from the day before. For some of the questions old and new, cue words had occurred during the letter-string tasks.

This complicated design yielded a simple finding: The clues helped. People were more likely to answer new and old questions alike when a relevant clue word occurred in the letter string task. Where there were no clues, subjects did not answer the old questions any better the second time around. Moreover—and this is where the Zeigarnik effect comes in—the clue words helped on the old questions more than the new ones. When subjects were primed by the first occurrence of a question, they were more likely to benefit from the clue than when the question only came up on the final round.

All this suggests that the initial impasse reached on some of the old questions primed a pattern in subjects' brains for detecting relevant information. The subjects may or may not have consciously noticed answers among the letter strings. Research from other studies suggests that often people do not become aware of such things, a matter not investigated in this study.

Seifert and her colleagues conducted a further study that directly addressed recall for problems. The investigators asked participants to tackle a number of puzzle problems that could be solved in anything from half a minute to several minutes. After the problem-solving phase, the participants simply tried to remember all the problems they could, writing out what the problems were. The investigators hoped to show that the participants remembered better the problems where they had reached an impasse. To test this question, the investigators manipulated the problem-solving phase in various ways. Some participants were interrupted on half the problems at random after about 30 seconds. Other subjects could work for the full 50 minutes allocated however they wished, completing many problems while leaving others at an impasse. Still others had exactly 1 minute to work on each problem. Comparing memory for problems across these conditions, the investigators concluded that unsolved problems left in a state of impasse were somewhat more likely to be remembered.

I am skeptical about whether Seifert and her colleagues constructed a compelling demonstration of the full-blown Zeigarnik effect here. The memory advantage for problem solvers at an impasse was not great. It is not clear that the experimental design really controls for the competing hypothesis of time-on-task—the longer a person works on the problem, the better it is likely to be remembered. Finally, a partial replication with more difficult problems yielded no edge for problems at impasse.

However, these concerns do not matter so much. Whether problems left at impasse are better remembered or not, the point still remains that problems tend to be remembered simply by virtue of being worked on. Maybe failure indices get established all the way along, not just at impasses. Maybe it is not just failure indices but relevance indices that the mind sets up, so that later chance encounters remind one of a problem already solved and create opportunities to extend or revise the solution. One way or another, the incidental learning that occurs while simply working on a problem primes patterns in the mind for later triggering.

Knowledge of the Genre of
Insight Puzzles

Agatha Christie's mystery *The Mousetrap* is London's longest running play—almost half a century on stage. When *The Mousetrap* came to Boston, my wife and I and another couple attended and found the play good fun. We enjoyed it all the way through, but by the middle of the evening, we were confident that we knew who the murderer was.

Identifying the murderer called for an insightlike shift of focus, but how did we arrive at this shift? Was our insight due to detailed consideration of physical evidence or detection of a subtle inconsistency in a character's alibi? Nothing of the sort. We figured that the culprit was most likely to be the one person who seemed least likely. And we were right.

The "least likely" principle is a good one for many mysteries. It goes hand in hand with a couple of other similar principles. The culprit is rarely anyone for whom the plot builds genuine sympathy, else audiences would feel regretful. Also, the culprit is rarely an extremely minor character, because it would not be interesting to discover that the milkman who had appeared once in the doorway was the murderer.

Knowing the genre of a problem and taking advantage of it is a powerful way of thinking. However, it is a very different way of thinking from working on the problem as such. This point holds for insight puzzles just as much as for other genres of problems. Perhaps one can know too much about them for them to require much insight.

I know the genre of insight puzzles well, so I decided as an experiment to examine how I used knowledge of the genre to solve the next several insight puzzles I encountered. This was an informal test of the idea that one can know too much for a problem to function fully as an insight problem. It is also another opportunity for readers to practice their breakthrough thinking on the following puzzles, the ones I happened upon next.

🖉 Pen Pals

Describe how you can put twenty-seven animals into four pens so that there is an odd number of animals in each pen.

🖉 Basketball

A basketball team won 72–49, and yet not one man on the team scored as much as a single point. How could this be?

🖉 The Lazy Policeman

A woman did not have her driver's license with her. She failed to stop at a railroad crossing, then ignored a one-way traffic sign and traveled three blocks in the wrong direction down the one-way street. All this was observed by a police officer, yet he made no effort to arrest the woman. Why?

🖉 Runaway Sheep

A farmer has seventeen sheep in a pen. All but nine escape. How many are left?

🖉 The Price of Wine

A bottle of wine costs $10. The wine is worth $9 more than the bottle. How much is the wine worth?

Solutions

I have seen many problems like these, both because of writing this book and before. Their familiarity helps me to solve them—but not always. Let me comment on each problem in turn.

Pen Pals. *As soon as I read this, I wondered whether the four pens had to be separate from one another. The assumption of separate pens could be a canyon trap. The puzzle reminded me of the Ten Matches problem, where the solution involved placing a smaller square in one corner of a larger square. Therefore, I imagined one big pen containing three little pens. It's easy to distribute the animals then—say, 7, 7, and 13. Each small pen contains an odd number, and the big pen contains all twenty-seven, also an odd number.*

I did not even bother to ponder whether separate pens were possible until after the above sequence of thought. They are not. Two odd numbers sum to an even number, so if odd numbers of animals occupy four separate pens, the sum has to be even.

Another even simpler solution puts the animals inside a pen inside a pen inside a pen inside a pen. This solution fits the stipulations of the problem, but it seems less plausible than one pen containing three pens. What animals would need as much penning as that?

Basketball. *My first thought was that the problem meant to trap me with a play on words—"not a single point." Perhaps the basketball players each scored much more than a single point. However, a quick look at the problem statement showed that this reading was anticipated and excluded by the phrase "not one man on the team scored as much as a single point." So, I asked myself, what else could be misleading, since obviously the players had to score? Gender occurred to me as a possibility. "Not one man," the problem said.*

Probably it was a women's basketball team. *This came easily because there are several other insight puzzles that turn specifically on gender stereotypes. They evoke situations where our default expectation is for males.*

The Lazy Policeman. *Here again I was alert for an easy canyon assumption that could mislead. In fact, the solution came to me while reading the second sentence. The title telegraphed that the policeman was not going to arrest the woman for some reason. In the odd world of insight puzzles, chances are the woman had not broken the law at all. How could the woman act as described without breaking the law? Perhaps she was not driving but walking. Nothing in the problem said that she was in an automobile.*

Runaway Sheep. *This problem I recognized right away as likely to contain an oasis of false promise. It tempted an immediate "subtract" response with the last sentence, "How many are left?" So I looked more carefully. On a careful reading, I discovered the key phrase "All but nine escape," so of course nine are left, no subtraction needed.*

The Price of Wine. *I got this problem wrong. Alert as in the previous problem for an oasis of false promise, I thought I had discovered one. The problem said, "A bottle of wine costs $10. The wine was worth nine dollars more than the bottle." I'm being tempted to subtract and answer "$1," I thought. I set out to read more carefully. If the wine is worth nine dollars more than the bottle, it's that "more" I should answer—$9.*

However, I did not check my answer carefully. If the wine is worth $9 and the bottle $1 for a total of $10, then

the wine is worth eight dollars more than the bottle,
which is not what the problem stipulates. The right
answer is $9.50 for the wine and $.50 for the bottle. Then
the total is $10 and the difference is $9 as required.
Why did I mishandle this simple problem? Because I
thought I had found the trap, but the real trap lay
behind the trap I saw. Why did I get the others right?
Partly because I was thinking with Klondike logic, with
a special eye out for the deliberate deceptiveness char-
acteristic of many insight problems.

Three morals can be drawn from this exploration of the genre of insight problems. First of all, it's advantageous to know the genre, just as knowing how mysteries are constructed gives one an edge in detecting the villain of *The Mouse Trap*. Insight puzzles are more approachable for the problem solver who is alert to language that creates canyons and oases—for instance, aware that gender stereotypes may be in play, wary of reflexive responses to key phrases, cautious about assumptions that things are separate when they might be combined, and so on. In the same spirit, it's useful to have a repertoire of known insight puzzles for making analogies.

The second moral is that knowledge of the genre can sometimes be entrapping. Use of such knowledge may create the illusion of having the correct solution, but one seeming trap may hide another.

Third and most important, when problem solvers exercise their knowledge of the genre of insight puzzles, they are deploying a kind of a niche skill, not full-scale breakthrough thinking. Insight puzzles are a good gymnasium for insight in general only to the extent that they remain fresh and varied, so that their tricks are hard to anticipate.

12

Knowing Too Much and Forgetting Enough

To Fly Again

The role of knowledge in insight is complicated. Sometimes an advantage comes from knowing facts that others do not and having the knowledge more in the foreground than in the background of one's mind. Sometimes this takes the form of having the right frame of reference—a lawyer, a plumber, a poet, a stockbroker. But the impact of knowledge can be double-edged. There is such a thing as knowing too much. And sometimes, you can know even more than too much, so much more that the trap set by the "too much" doesn't matter. Remember The Fly. This problem from an early chapter invited us to imagine two people walking slowly and steadily toward each other from 10 feet apart at a rate of 1 foot every 10 seconds. A fly flies back and forth between their noses at 1 foot per second. How far does the fly travel before getting crushed between their noses?

The way the problem is put suggests somehow summing up the decreasing nose-to-nose distance of each trip the fly makes as the two people slowly converge. This is in fact an infinite series of progressively smaller distances. There are many technical strategies for deriving the sum of an infinite series. The problem sends a strong invitation to mathematicians to deploy one of these strategies. A person with less knowledge might or might not solve the problem, but would not fall into that oasis of false promise.

However, if the mathematician is skilled enough, the trap may not matter. The story is told of someone presenting The Fly to the brilliant mathematician John von Neumann. In a matter of seconds, von Neumann correctly answered, "50 feet."

"Oh," said the puzzler, "You saw the trick."

"What trick?" said Von Neumann. He had simply done the complex mathematics of the infinite series in his head.

This story probably is not true. Indeed, I have found that sometimes the same story is told both of von Neumann and of the prodigious Massachusetts Institute of Technology mathematician Norbert Wiener. Both mathematicians could do such feats, but did either one actually face The Fly and respond in this way? Whatever the truth of the matter, the story makes an important point. Knowledge is paradoxically both empowering and entrapping. It all depends on the problem. And sometimes enough expertise can even get one past a trap.

Knowing Too Much for Your Own Good

In the play of the mind, ignorance is usually a villain. It is not good not to know. Information is the fuel for the mind's explorations. But anecdotes such as the one about von Neumann suggest that one can also know too much. Is there any systematic evidence of this?

A classic concept in psychology speaks directly to the question. *Einstellung* is a German term that can be translated as "mental set." Problems that invite a particular approach can establish a mental set that exacerbates the difficulty of seemingly similar problems that require a shift in approach.

There are many demonstrations of the *Einstellung* effect in the psychological literature, using puzzle problems of one sort or another. Steven Smith, a psychologist at Texas A & M University, and David Jansson, a mechanical engineer, stepped away from puzzle problems to examine mental set in the context of naturalistic invention. They posed creative design tasks to design engineers. The tasks included sketching designs for a bicycle rack, a measuring cup for the blind, a device for taking readings inside intestines, and a disposable spill-proof coffee cup. The latter task went something like this:

✎ A Disposable Spill-Proof Coffee Cup
Design a disposable spill-proof coffee cup. The illustration shows one way this might be done. However, bear in mind that you had best not use straws or mouthpieces in your design. These can prevent cooling and cause scalding.

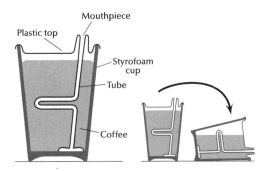

As usual, readers may want to attempt this design problem before continuing.

Smith and Jansson actually provided more information than is helpful for the Disposable Spill-Proof Coffee Cup problem. The illustration hinders rather than helps. Half their subjects received the coffee cup illustration and half did not. With an illustration at hand, people's creative efforts tended to echo it, showing much less diversity and inventiveness. Moreover, the designers often ignored direct instructions to be cautious about features of the illustrated designs. For the coffee cup, 56 percent of the designs done by subjects who saw the illustration included a straw or mouthpiece, despite the explicit caution. In contrast, only 11 percent of the subjects who did not see the illustration incorporated these features.

Certainly, looking at what others have done can often be a huge help. However, it can also generate an unwelcome fixation. An illustration creates an oasis trap when it displays seemingly attractive features that prove hard to move away from.

Fertile Forgetting

So what to do? Maybe just take a break. It stands to reason that a person might produce a more creative cup design after moving away from the problem, because the details of the illustration might fade, leaving the person freer to think more flexibly.

Smith and Jansson did not pursue such a study. However, with other kinds of problems, Smith directly investigated fertile forgetting. One kind of problem Smith used might be called a *word rebus*. Rebuses, recall, are made of images and sometimes words that together make a phrase. A rebus for "fly ball" would be a picture of a fly and a picture of a ball. A word rebus adopts the same idea but without pictures, presenting a layout of words where the arrangement suggests a well-known saying. Here is an example:

✎ Word Rebus Example

you   just   me

This word rebus means "just between you and me."
With "just between you and me" as a model, here are three more word rebuses from Smith.

✎ Word Rebus 1

r | e | a | d | i | n | g

✎ Word Rebus 2

$$\frac{0}{\text{B. A.   M.D.   Ph.D.}}$$

✎ Word Rebus 3

Wheather

If these prove difficult, here are two rounds of clues. (Mask the right column to see only the first round to start with.) However, be warned that in some cases the clues may be misleading.

Word Rebus	First Clue	Second Clue
1	between	lines
2	below	degrees
3	under	not

Smith and his colleague Steven Blankenship conducted a formal study using word rebuses such as these to investigate fertile forgetting. In their initial presentation, some rebuses came with helpful clues and some with misleading clues. After their first-round efforts, subjects were retested on the rebuses they did not get, but without the clues. Subjects first tried to solve an unsolved rebus and then to remember the clues. Some were retested right away, some after a delay. Smith and Blankenship found that longer delays yielded poorer memory for the clues and improved problem solving for the rebuses with misleading clues. Smith has repeated this work with two other kinds of problems, with the same results. His data demonstrate that *fertile forgetting* indeed can make insights accessible that otherwise would be blocked. Smith concludes that fertile forgetting is a major factor in incubation.

It may seem odd that the experimenters in this study deliberately included misleading clues in some items. However, remember that the world often presents problems that mislead. To include such clues systematically in an experiment is just a way of recreating in the laboratory the blocks that too often hobble progress in realistic settings.

Solutions

As to the numbered rebuses, the solutions are as follows: (1) "reading between the lines"; (2) "three degrees below zero"; (3) "a bad spell of weather."

The clues to the first two problems are helpful, but the clues to the third are misleading. In particular, "under" creates an under-the-weather association with the word "weather," which has no relation to the solution. Likewise, "not" creates a whether-or-not association with "whether," also unrelated to the solution. Smith notes that the third problem also might prove harder because of an *Einstellung* effect: The example and the first two problems all depend on relative position to generate the target phrase. However, the last problem has nothing to do with relative position. It's a play on spelling.

Unforgettable

Fertile forgetting is a nice notion. But what if you can't forget? What if the knowledge in play is not some recent acquisition from the problem situation itself but thoroughly familiar knowledge from years of experience? An interesting demonstration of this dilemma comes from the following task.

✐ Alien

Many a youngster has spent considerable quality time in elementary school drawing space aliens and other interesting images during English or Arithmetic. This is your task now. Sketch an alien from outer space. Make the alien as alien as possible, striving for something truly strange.

A similar task, devised by Tom Ward of Texas A & M, aimed to explore how original people could be when asked to draw an animal from another planet. Ward administered the task to hundreds of college students in a series of experiments. He examined the results to appraise how well the participants managed to break the mold.

Interestingly, the kinds of animals the students drew proved to be remarkably predictable, not in the details but in overall configuration. Most were bilaterally symmetric. Most had at least one familiar sensory organ—eyes, ears, or nose. Most had at least one familiar kind of major appendage—arms, wings, or legs. Readers who attempted this puzzle will probably find such features in their sketches. The nearby figure offers two examples from Ward's research. The birdlike figure was one produced by a student from a group asked to draw a creature with feathers. Notice how the student has incorporated a number of avian features besides feathers, even though these were not specified.

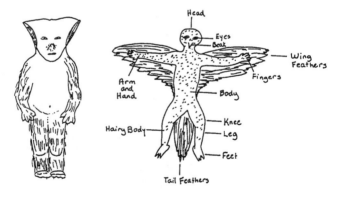

Ward interprets his results to say that people find it difficult to escape from templates well established in their minds. Features like eyes and bilateral symmetry are deeply ingrained because of the appearances of animals on Earth. Indeed, the students reported relying more on their knowledge of Earth animals than on that of fictional alien creatures. While fertile forgetting may provide a useful mechanism of incubation in some ways, it is not likely to help problem solvers to escape from deeply entrenched patterns. In Klondike terms, entrenched patterns like the tacit mental models of the standard alien amount to narrow canyons of exploration. People easily move within the boundaries of their alien canyon, spinning variations on head shape or skin texture or exact form of eyes. However, transcending those boundaries does not come readily.

Many insight puzzles depend on unforgettable blocks. Recall The Polygamist, about the man who married ten women. No one is likely to forget the most prominent meaning of *married*—"getting married to"—which gets in the way of the secondary meaning, "marrying one person to another." Also, consider the problem Pen Pals, about how to group twenty-seven animals into four pens. No one is likely to forget that pens normally stand separate from one another. Remember Basketball, the game in which no man on the team scored as much as a single point. People cannot easily forget their gender stereotypes. Such cases as these teach that fertile forgetting is an important but limited mechanism for navigating Klondike hazards. Too much is unforgettable. While just waiting may help sometimes, the active use of breakthrough thinking to break a *mental set* (a fixed idea or attitude) remains key.

Klondike Psychology

Time to take stock. The last several chapters have explored how the human mind works during breakthrough thinking. The very question would disappear if breakthrough thinking were no more than normal reasonable problem solving—if, for example, major scientific discoveries that emerged suddenly were simply the consequence of especially astute and informed reasoning. But despite the studies of Langley, Simon, Bradshaw, and Zytkow, it was argued that breakthrough thinking had its own unique character, governed by a search through the Klondike possibility spaces of unreasonable problems.

But what was distinctive about breakthrough thinking from a psychological standpoint? The vivid experience of sudden insight has made it natural to propose mental overdrives—specialized psychological processes that produce cognitive snaps. However, upon examination, the existence of these mental overdrives is suspect. Such proposals can be viewed critically as products of unwarranted hypostatization—ascribing distinctive material existence to something without compelling evidence.

With mental overdrives dismissed, we can better understand breakthrough thinking as a process of searching through a Klondike possibility space that sets the problem solver up for a sudden resolution. The consummation occurs either through an episode of reasoning much shorter than the overall search or by the rapid falling-into-place characteristic of ordinary understanding.

However, none of this means that breakthrough thinking is business as usual. Three psychological aspects of breakthrough thinking loom large: active knowledge, pattern priming, and breaking mental sets. As to the first, breakthroughs inevitably depend on knowledge, but not just knowledge buried in the mind's basement. A great deal of what we know is inert, far away from the forefront of our thinking. Breakthrough thinking requires not just possession of, but ready access to, the right knowledge.

Pattern priming names the mental readiness that develops when struggling with a problem has saturated us in it. Somehow, the effort and investment set up mental watchdogs that are likely to bark when a relevant clue comes along—even if the problem solver is in a different situation altogether.

Finally, having a mental set amounts to pattern priming in a negative direction. We often bring to a problem, or develop during early work on it, implicit assumptions that entrap our thinking. Our minds are again primed, but in an unhelpful rather than a helpful way. Breakthrough thinking involves breaking through such mental sets.

Active knowledge, pattern priming, and breaking mental sets are part of what might be called not Klondike logic but Klondike psychology. Klondike psychology has to do with what's difficult about breakthrough thinking for the human mind, and how the human mind sometimes dodges those difficulties. But what does all this mean for the art of breakthrough thinking? An illuminating answer involves a look back at the four characteristic challenges of breakthrough thinking and responses to them: wildernesses of possibility and roving, clueless plateaus and detecting hidden clues, narrow canyons of exploration and reframing the situation, oases of false promise and decentering from false promise.

Roving by the active use of varied knowledge. Imagine Sherlock Holmes bringing to bear a fund of knowledge while investigating the scene of a crime: types of cigarette ash, the nature of soil in different parts of London, current political intrigues, previous crimes he's solved, chemical analyses that could be applied. In considerable part, roving around a wilderness involves activating the varied knowledge one has and putting it to work. It may be helpful to summon knowledge of prior related problems if that knowledge does not come forward naturally. It may be helpful to use analogies and metaphors to connect the problem at hand to other areas—a kind of selective combination and/or comparison, to use the terms of Davidson and Sternberg. It may be helpful to look at the problem through the perspective of another viewpoint or discipline—deliberate bisociation. Such moves as these rove around the Klondike possibility space in a broad and flexible way by aggressively marshaling available knowledge, although the exploration may be limited by any of the following traps.

Detecting hidden clues by pattern priming. To continue with Holmes, imagine the man immersing himself in an investigation with characteristic obsession, living with it all day and all night, getting primed to catch the hidden clues that tell the story. One way of detecting hidden clues is through saturation in the problem to create pattern priming, coupled with close inspection to feed information to those mental watchdogs that get set up. The clueless landscape of a plateau may then dissolve into a more meaningful and suggestive landscape. Time away from the problem may be helpful; one may encounter the needed clue elsewhere.

Reframing the situation by breaking mental sets. Holmes warned Watson repeatedly against making early assumptions, held himself open to the evidence, and remained ready to change his view of the crime radically as the evidence told more of the story. Virtually by definition, the assumptions that lock one into a particular view of a problem are mental sets or frames. Focusing on different combinations of clues and constraints, adding what was excluded, and excluding what was included may help to break one's mental set (deliberate selective encoding, one might say). A change in representation brought on by importing other concepts or frames of reference can help to break set (deliberate selective combination or bisociation). So can fertile forgetting, brought about simply by staying away from the problem for a while.

Decentering from the false promise of beguiling solutions. Sherlock Holmes was forever skeptical of the evident solution, the apparent perpetrator, the obvious modus operandi. Recall that getting away means getting away from oases—intractable near-solutions and mediocre solutions. Beguiling solutions generate the ultimate mental set. To break this mental set, it's generally necessary to notice the fixation, deliberately set aside the solution, and insist on looking elsewhere, at least for a while.

A Klondike Psychology Playground

Here are several insight puzzles to work on. They provide plenty of opportunities to muster a wide range of everyday knowledge, to get your patterns primed and look for clues, and to break mental set. Get around, get sharp, get beyond, and get away!

✐ Marriage Law
Should a man be allowed to marry his widow's sister? Why or why not?

✐ The Apple Orchard
Johnny Appleseed decides how to organize his next planting of ten apple trees. He wants them in five rows, with four trees in each row. How does he do it?

✐ The Card Players

Four people are seated at a table, playing cards. Yet all four lose. No other people are present. How can this be?

✐ The Cigarette Maker

A bum discovers that he can glue together five cigarette butts to make one cigarette. Through assiduous effort, he scrounges up twenty-one cigarette butts. How many full smokes does he get?

✐ The Rope Ladder

A rope ladder hangs over the side of a boat near a dock. The fifth rung is just under water. The tide is rising at a steady rate of 1 foot per hour. Each rung is $\frac{3}{4}$ of an inch thick, and the gap from one rung to the next is 8 inches. How many rungs will be under water in 3 hours?

✐ Household Tasks

Betsy and Boris have three chores to perform around the house: (1) Vacuum the floors. They have only one vacuum. The task takes 30 minutes. (2) Mow the lawn. They have only one mower. The task takes 30 minutes. (3) Feed and bathe the baby. This also takes 30 minutes.

How can Betsy and Boris work together to get the chores done as early as possible?

Solutions

Here are the solutions with comments.

Marriage Law. *No, the man should not. If he has a widow, that means he himself is dead. Why is it easy to* miss this anomaly? *The reason is a narrow canyon of exploration created by an encoding problem. People tend to assemble cues into a sensible pattern, so on hearing this problem they interpret the scenario as one where the man's wife has died: Should he be allowed to marry her sister? However, this is not what the problem asks.*

The Apple Orchard. *The conditions can be met by arranging the trees in a star shape. This problem can prove difficult because of a canyon trap: People are used to thinking of rows as parallel to one another and it's hard to break that* Einstellung.

The Card Players. *The people are each playing solitaire. It is natural to encode the situation as four people playing cards with one another. However, the problem statement only says they were seated at the same table and playing cards.*

The Cigarette Maker. *The obvious answer, an oasis of false promise, is four cigarettes, with one butt left over. However, after the bum smokes the four, that yields four more butts. With the leftover butt, the bum now has five butts to make another cigarette. Therefore, the bum gets five smokes—and still has a butt left over after he finishes his last smoke.*

The Rope Ladder. *In this puzzle an encoding problem sets up an oasis of false promise. The tendency is to encode the situation as one where the water rises over the rope ladder. However, notice that the ladder is attached to the boat. This means that the boat and the ladder rise with the tide. Therefore, the number of rungs under water in 3 hours is still five.*

Household Tasks. *There is a natural-oasis solution: 60 minutes. However, there is a better selective combination. If Betsy and Boris simply divide up the tasks—a canyon assumption—either Betsy or Boris will have to wait while the other finishes the last task. The trick is to keep Betsy and Boris both working all the time. This can happen if they share tasks. Betsy might do half the vacuuming (15 minutes) and all of the baby care (30 minutes). Meanwhile, Boris would start with the lawn (30 minutes) and then finish the vacuuming (15 minutes). Total time: 45 minutes.*

Part 4

Does Nature Think?

In which we return to the comparison between human breakthrough thinking and biological evolution, asking how a blind process like evolution accomplishes breakthroughs, exploring how cultural evolution sometimes works like biological evolution, and examining why the need for breakthroughs is ubiquitous and inevitable.

13

Evolution Breaks Through

Ten-Million-Year Snaps

How long does a snap of discovery take? From a second to 10 million years. This last figure clocks in at about 10^{13}, or 10,000,000,000,000 times as long as the one-second end of the spectrum.

What kind of a snap could possibly take as long as that? The answer is an evolutionary breakthrough. The blind process of evolution somehow invents organisms, including bumblebees, orchids, duck-billed platypuses, and brainy creatures such as ourselves. And from time to time the process of evolution achieves fundamental breakthroughs. One was the emergence of complex multicellular organisms about 650 million years ago. Another was the development of flight—not just once but several times, for instance in insects, birds, bats, and flying reptiles.

The first chapter of this book suggested that the basic story line of breakthrough thinking occurs in other settings besides the human mind. One of them is biological evolution. Bizarre though this idea may seem, it suits the Klondike model of breakthrough thinking and its central tenet: Breakthroughs are not fundamentally psychological but structural phenomena. The dynamics of breakthroughs reflects not the way the human mind works specifically but the general challenges of searching in possibility spaces with Klondike characteristics. Wherever such searches occur, so will the typical profile of breakthrough thinking—the long search, the precipitating event, and so on.

Let's rewrite that profile so that it makes no specific mention of human minds. Adapted from Chapter 1, here's what it sounds like:

1. *Long search*. A breakthrough event follows a long search.

2. *Little apparent progress*. The long search shows little apparent progress in the direction of interest (although it may take other interesting twists and turns).

3. *Precipitating event*. The breakthrough gets triggered by a precipitating event of some sort, a key step in the search or an external accidental occurrence.

4. *Rapid breakthrough*. After the precipitating event, the process quickly converges to yield the breakthrough itself. This is the generalized equivalent of the cognitive snap.

5. *Transformation*. The breakthrough transforms things in a generative way. It is surprising relative to the situation before, representing a distinctly fresh direction.

Phrased this way, the features of a breakthrough event are stripped of mentalism. There is no reference to thinking or intelligence or understanding. As the case of evolution suggests, such Klondike searches can occur by entirely blind and mindless mechanisms. But the basic pattern still holds. At least, that is the case to be made.

The Breakthrough of Birds

The emergence of birds is one of many evolutionary stories of success snatched from the jaws of defeat. The losing jaws in this case were the dinosaurs, who departed this planet in a massive wave of extinction around 65 million years ago. It's now generally recognized that birds descended from the dinosaurs, who therefore live on in aerial form. Some scholars have even proposed that the taxonomy of animals be rewritten to make birds a kind of dinosaur.

Evolution has been at work for almost 4 billion years of the Earth's 4.5-billion-year history. In all that plethora of biological change, evolution's discovery of birds is particularly easy to relate to, both because birds fill our skies and because flight has figured centrally in human quests. The basic time line goes something like this:

Millions of
Years Ago

225	Emergence of the dinosaurs.
200	Emergence of pterosaurs (pterodactyls), flying lizards.
175	Earliest fossil birds, Protoavis, already with fully developed feathers. Bird fossils from this era are rare. Pterosaurs continue to dominate the skies.
150	Archaeopteryx, the other known early fossil bird. Pterosaurs still thrive.
95	Appearance of water birds similar to modern type.

| 65 | Extinction of the dinosaurs, pterosaurs, and many other forms of life. |
| 35 | Dramatic radiation of birds since the demise of the dinosaurs, including flightless birds, many modern types. Robert Wesson in his 1991 *Beyond Natural Selection* notes that "the profusion of specialized birds left the pterosaurs far behind." |

How well does this ladder of evolution fit the profile of a breakthrough event?

Long search. Evolution's search for viable forms has persisted for several billion years and continues all the time. It is not, of course, an intentional search guided by any agent. The process of variation and selection by survival of the fittest happens automatically.

Little apparent progress. The idea of progress has to be used cautiously in connection with evolution. Stephen Jay Gould in his *Wonderful Life* warns that evolution does not seek higher forms like primates but simply sprawls in adaptive directions. However, if progress means progress toward filling a particular adaptive niche like medium-sized flying organisms, then one can ask whether evolution advances steadily on the front of flight. The answer is no. Reading across the fossil record, birds emerged relatively suddenly in geological time, without a long history of immediate bird precursors.

Precipitating event. This can only be speculative, since the early fossil record of birds is sparse. As the chronology above indicates, there seem to have been two critical junctures: the development of feathers and at least partial flight around 200 million years ago, and the rapid adaptive radiation of birds some 50 million years ago.

As to the first, the precipitating event presumably was the accidental appropriateness of feathers for supporting partial flight, perhaps in the form of hopping and gliding. Preflight feathers may have been insulation for warm-blooded dinosaurs, or brightly colored feathers along the "arms" of dinosaurs may have served as sexual displays, a pattern still evident in many modern birds. Whatever the source, their occurrence set the stage for rapid evolution toward flight.

As to the second precipitating event, the natural hypothesis looks to typical evolutionary opportunism: The birds were filling the ecological niches abandoned by the extinct pterosaurs.

Rapid breakthrough. Again the story here has to take into account the two-phase evolution of birds indicated in the chronology. Feathers appeared early in the fossil record, "out of the blue." The earliest feathers anatomically have virtually the same structure as the feathers of modern birds. This suggests that feathers were rather rapidly achieved.

While feathers may have evolved quickly, birds were not a conspicuously successful species early on. Some 150 million years lay between the first birds and the great diversification of birds, which in a period of some 30 million years yielded a range of species something like those that thrive today. This, then, was the second geologically sudden event in the development of birds.

Transformation. Certainly flight represents a radical departure from the initial adaptations of those organisms who have achieved true wings, all of them initially ground dwellers. According to Robert Wesson, the especially distinctive achievement of birds was feathers, which are complex organs with intricate interlocking structures that keep them light and stiff. Feathers serve the purpose of insulation as well as flight, but furlike structures have evolved many times. Batlike wings have evolved three times (pterosaurs and the two kinds of bats), feathered wings only once.

The bird breakthrough event yielded the entire range of birds in all their variety, with adaptations to a wide range of very different ecological niches. Gould notes that large ground birds even assumed the roles of major predators, at least as successful as marsupials, in South America before it became connected to North America a few million years ago.

Evolution as a Kind of Search

Some episodes of evolution, including the emergence of flight, follow the pattern of a breakthrough event. Why should this be? Because the same Klondike challenges that explain breakthrough thinking by humans also explain breakthrough events in evolution. Understanding the details requires describing evolution as a process of smart search in a space of possibilities.

Consider as an example how dinosaurs perhaps took flight. Only decades ago, it was generally believed that dinosaurs were cold-blooded, but today paleontologists think it likely that many dinosaurs were warm-blooded. This characteristic helped them evade the torpid behavior of normal cold-blooded reptiles in cold weather and darkness, keeping the biochemical engines turning over at a steady pace and the organism active as with mammals today. Body heat regulation is the principal adaptive challenge of warm-bloodedness. Mammals solve this problem with hair to insulate and sweat to cool off, among other means. In one account of the evolution of birds, feathers did not evolve initially to serve flying. They were certain dinosaurs' answer to the problem of keeping warm, as much insulation then as for birds today.

What happened then could have been something like this: Many dinosaurs walked on two legs, leaving two free for grasping. Feathers along the two "forelimbs" may have afforded an adaptive advantage in giving balance. Through natural selection the feathers lengthened. By chance, the longer feathers afforded some lift and glide, at first perhaps just allowing for slightly longer jumps, then, as the feathers lengthened still more, for true glides and eventually true flight. According to this account, feathers along the fore-limbs progressed through four roles: first for insulation, next for balance, then for glides, and finally for flight, during each phase an adaptive advantage that drove evolution forward until dinosaurs filled the air.

We can envision this process of evolution as searching in a fitness landscape of different biological forms that might or might not prove viable. The accompanying figure offers a graphic illustration of what this might be like. Concentrate for now on Fluffasaurs, Glidasaurs, and True Birds, ignoring Wonderbirds and Superbirds until later. Imagine a thriving population of Fluffasaurs, small dinosaurs with insulating feathers. The hope is that evolution will make it by way of Glidasaurs to True Birds. To do so, evolution has to navigate a narrow path. The diagram suggests that organisms along the path are only marginal survivors, "barely viable."

To appreciate the challenge, imagine a piece of this land-

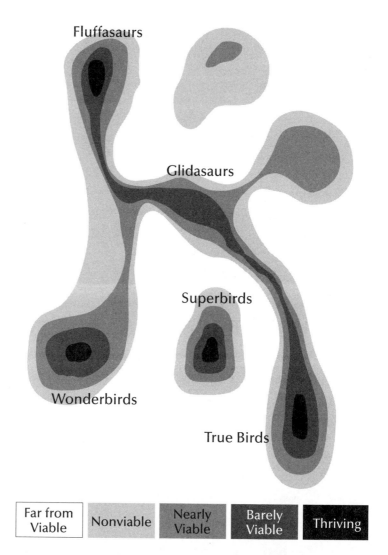

scape of possibilities centered around the Fluffasaurs. Here are just a few of the dimensions of this space relevant to insulation, stability in running, and flight:

- Total quantity of feathers
- Length and stiffness of feathers
- Distribution of stiff, long feathers: on the forelimbs versus elsewhere
- Total mass

According to Darwin, evolution explores these and other dimensions by a process of variation, selection, and preservation of traits:

> *Blind variation.* Young organisms show variations relative to one another and their parents—a little taller, shorter, quicker, slower; having thicker hair, thinner hair, etc. These variations are blind trial and error: They may or may not help the organism to survive. Today we understand this as reflecting variations in DNA due to mutation and the gene shuffling of sexual reproduction, but Darwin did not know about DNA.

> *Selection through survival and reproduction.* Variations that yield an adaptive advantage help the organism to survive long enough to breed and produce offspring. The effect may be small—a mere statistical edge—but over time it becomes cumulative.

> *Preservation through inheritance.* Variation notwithstanding, offspring tend to preserve the features of their parents. Both adaptive and maladaptive traits are passed on by inheritance, but organisms with adaptive traits are more likely to survive and reproduce, so those traits appear with greater frequency in the next generation.

How does this translate into the world of Fluffasaurs? Each generation yields a range of variation, Fluffasaurs with more or fewer feathers, stiffer feathers or less so, differently distributed feathers, and greater or lesser body mass. The more viable forms survive and breed, and the next generation shows more of their survival characteristics. Contemporary views of evolution, such as the punctuated equilibrium theory mentioned in Chapter 1, elaborate the implications of this basic Darwinian mechanism, but by and large they do not challenge it. (However, in the next chapter we'll look at some revisionary ideas about evolution, and their Klondike implications.)

The most important thing to recognize about this search process is its blind character. Evolution does not aim in the direction of true birds. Evolution does not aim at all. Rather, trial-and-error variations occur. Variant organisms sprawl out in all directions from the Fluffasaurs. Survival and reproduction determine which directions thrive and which falter and fail.

Of course, this is a simplified and speculative example. It expresses a particular account of how birds may have evolved—by way of insulation feathers—and what some of the critical dimensions of adaptation might have been. Also, organisms do not evolve by themselves. Other organisms—Fluffasaur prey and Fluffasaur predators, as well as those less directly involved in Fluffasaur life—are evolving at the same time, complicating the dynamics. The viability of an evolving organism varies not only with its features but also with the evolving characteristics of other organisms in the same ecosystem. However, it really does not matter whether the story of Fluffasaurs as told here is just right, so long as it provides a plausible example for discussing the Klondike structure of the possibility space and the breakthrough logic of evolution.

The Breakthrough Logic of Evolution

Although blind, evolution's search for new more adaptive forms of life is not stupid. Characteristics of the process are well suited to deal with the four Klondike challenges of wilderness, plateau, canyon, and oasis traps. Like human creative thinking at its best, evolution involves features that "up luck," increasing the chances of doing well in the unreasonable chaos of a Klondike space. Of course, evolution involves strategies somewhat different from the Klondike logic of human beings. How this works can be seen through a closer look at the possibility space of Fluffasaurs.

A wilderness of possibilities. The wilderness trap recognizes that in many search situations, there are innumerable candidate solutions but very few genuine solutions. In the case of evolution, the wilderness trap takes a simple form: Very few candidate organisms are actually viable. In the diagram shown earlier in the chapter, only the darkest and next darkest regions represent viable organisms amid a sea of nonviable dinosaur-like structures.

To see why viability is rare, imagine variant organisms generated by just the four factors mentioned earlier— quantity of feathers, length and stiffness, distribution, and total mass of the organism. Suppose a gene shuffler randomly twiddles these factors to concoct new Fluffasaurs. How many of these mixes could survive? Consider the hazards: A lot more feathers, and they would be a heavy biological investment to grow and lug around. A lot fewer, and they would neither provide well for insulation nor lead toward flight. Longer stiffer feathers might yield creatures more likely to evolve flight, but at the cost of insulation, so the creatures might die of cold. Differently distributed feathers would also lead into tradeoffs between insulation and the prospects of flight. Feathers are needed out on the forelimbs for balance, gliding, and flight, but around the Fluffasaur body for insulation.

As to mass, genetic variation cannot alter mass without encountering trouble on other fronts. A bigger Fluffasaur may enjoy some survival advantages due to size, but it needs more food and is less likely to find any advantage in gliding or flight, because it would need a much bigger investment in long stiff forelimb feathers to get off the ground. Notice how small birds have relatively short wings compared with body length, whereas large birds have much longer wings by the same measure. Basic physics comes into play here. The lift of a wing depends on its surface area, but the weight the wing has to carry depends on the volume of the bird. Scale up a bird by doubling its linear dimensions, and that makes eight times the weight to carry (2^3) but only about four times the lift (2^2). This will not do. Everything has to work together.

To generalize beyond Fluffasaurs, very few combinations of biological features external and internal would actually make up viable creatures. The possibility space of conceivable biological forms is huge, with only a scattered few who are fit enough to survive.

How, then, does Darwinian evolution deal with a wilderness of possibilities? By brute force. Evolution enjoys two great advantages: time and parallelism. Evolution operates over geological time. Even the punctuated equilibrium theory, which holds that evolution occurs in spurts, does not fundamentally challenge the point. The "spurts" still occupy innumerable breeding cycles and tens of thousands of years in multicellular organisms. As to parallelism, evolution tries many things at the same time. Each generation of an organism yields a range of variants. Each variant constitutes a parallel trial.

Clueless plateaus. Plateau traps are dilemmas of cluelessness: One direction of search looks about as promising or unpromising as another. First of all, what form does this plight take in the world of biological organisms? Many variations on the plan of an organism may be reasonably viable, with nothing to indicate which variation points in the most promising direction. This holds especially in a rich ecology, where several minor variations find places to thrive. How, then, can evolution tell that adaptation in the direction of flight would be especially fertile, when adaptation in a dozen other directions yields viable organisms too?

The answer is that evolution cannot tell, and does not need to. Evolution is blind. As mentioned earlier, evolution deals with plateaus through brute force, trying all directions in parallel, sprawling outward from the parent organisms. Evolution operates by filling the plateau, so to speak, rather than by choosing a direction along the plateau.

Narrow canyons of exploration. Canyon traps occur when a search process easily ranges within a certain region of the search space, but the breakthrough discoveries sit in another partially isolated region. The challenge is getting from one region to the next. In the case of evolution, partial isolation can be envisioned as a narrow path of marginal viability leading from the initial organism to a radically different form. For Fluffasaurs to evolve into true birds, evolution must pass along the path, with just the right balance of size, insulating feathers, pre-flight feathers, and so on. The path is narrow because at least two factors stand in competition: feathers as an investment in insulation, and feathers as an investment in partial flight. The wrong balance, and the intermediate organism will not be viable at all.

In the human case, not only narrow paths but many other factors contribute to canyon traps. People may perceive a problem in misleading ways, as many insight puzzles over the last pages have demonstrated. But evolution does not perceive anything. The canyons of evolution primarily reflect the narrow-path problem.

Canyon traps yield to the long-term parallel character of evolution. Rather than a mechanism searching for a path, think of a tenuous gas of organisms starting at Fluffasaurs and gradually diffusing to fill the dark region labeled "thriving" and then the slightly lighter region labeled "viable." With luck, some molecules of this gas spread along the thin "barely viable" neck toward true birds. Of course, the process is probabilistic. It might be that no molecules in this gas diffusion image of evolution wander into that narrow neck. Had that happened, there would be no birds.

This account of the breakthrough logic of evolution demonstrates its power. Searching in all directions with time and parallelism blows away the four Klondike challenges. Sort of.

But the diagram also illustrates a weakness of the evolutionary process. Paths of evolution must be paths of viable organisms, each one in its day at least marginally successful. Evolution can make it from Fluffasaurs to True Birds because a path of viability connects the two. However, consider what remains undiscovered. Suppose that Wonderbirds and Superbirds stand for two very effective birdlike biological designs. Evolution cannot make it to either one. Superbirds stand entirely isolated, an island in the white, utterly nonviable region. Wonderbirds seem to be a better bet because the diagram shows a connection from Fluffasaurs to Wonderbirds of bioforms that are "near viable." But this doesn't provide enough of a path for evolution to traverse. Only survival will do.

In this respect, Darwinian evolution proves much less flexible than human thought. Human creators commonly follow paths of mere promise, not even marginal viability, to see where they lead. The humans rove around, detect hidden clues, reframe situations—all of which can leap to new solutions. Indeed, human creators often leap across regions of no promise at all. Starting from Fluffasaurs, an Edison in the recombinant DNA workshop of the future might create Wonderbirds and even Superbirds. However, evolution cannot, so now there are none.

Oases of false promise. Oasis traps arise in problem-solving situations when the search process lingers near partial or merely adequate solutions, trying to improve them rather than exploring new turf. What this means here is that evolution explores variations more slowly as viability tapers off, and not at all in nonviable regions. Turning to the diagram again, evolution will generate many varieties of Fluffasaurs in that supportive black region called "thriving." However, in the marginal "barely viable" areas, the spread will occur much more slowly, and in the nonsurvival areas not at all.

Evolution's breakthrough logic for oasis traps is to spread into regions of marginal survival. It's important to recognize that evolution does not optimize, focusing only on organisms that thrive. It goes anywhere organisms can survive. Imagine the small Fluffasaurs, well adapted for a vigorous life as warm-blooded predators. Among many variants and variants of variants, genetic shuffling eventually produces a small dinosaur with stiff wing feathers that help with balance and long gliding hops. This Glidasaur might be less viable than Fluffasaurs. Nonetheless, it has its own niche, taking advantage of its better balance to feed on a flying insect population that evades other small dinosaurs. But there is a cost: The wing feathers impair the use of its forelimbs for defense, so it falls to larger predators more often. Its population expands and diversifies very slowly.

No matter. Evolution, which cannot care about anything, does not care that this intermediate form proves less well adapted than Fluffasaurs, its thriving ancestors. So long as it can barely survive, that suffices. In this sense, blind evolution has an advantage over human inventors. Whereas human inventors can be disturbed by the seemingly lower promise of a marginal variant, evolution cannot. Thus natural selection gradually strengthens and lengthens the feathers and supporting musculature of the Glidasaurs. With longer glides and partial flight, the creatures not only can feed better on aerial insects but escape their enemies. True Birds begin to emerge.

The Breakthrough of the Burgess Shale

The development of birds is a late chapter in the story of the evolution of multicellular life, volume 2 of "Life on Earth." Volume 1, concerning single-celled life, is by far the larger in time. Curiously, life in minimal form does not seem to be that difficult a chemical achievement. The earth is approximately 4.5 billion years old. As far back as 3.75 billion years there is evidence of life, which might even have arisen earlier—rocks older than this have been so transformed that they would not show the symptoms of life anyway. But while long, this first volume is not especially dramatic. For almost 2.5 billion years, the only life-forms were single cells of simple internal structure. More complex cells eventually developed and held the stage for another 700 million years.

One of the puzzles of the evolutionary record concerns why more complex multicellular forms of life took so long to appear. Apparently, the coordination of activities across cells to create larger organisms was a severe Klondike challenge for evolution. This challenge was finally met some 570 million years ago, at the beginning of the Cambrian era, in what is called the "Cambrian explosion." In the evolutionary blink of some 50 million years, a variety of complex multicellular life-forms appeared in the seas.

Our best snapshot of life in those remote days comes from fossil samples gathered in the Canadian Rockies at the eastern border of British Columbia. The fossil deposits of the Burgess Shale, named for the nearby Burgess Pass by the great naturalist C. D. Walcott, were formed when undersea landslides occurred. These landslides swept tons of silt, along with diverse organisms that lived in and above the silt, down into lower waters. There, low-oxygen conditions deferred decay while the chemistry of fossilization went to work. The consequence 530 million years later is rich fossil beds preserving in stone not only the hard, shell-like parts of those prehistoric organisms but the soft body parts as well. The summary time line looks like this:

Billions of
Years Ago

3.7   Chemical signs of life

1.4   Appearance of complex cells

0.570   Beginning of the Cambrian era, the

"Cambrian explosion" of diversification of multicellular life

0.530   Burgess Shale organisms. More recent findings at other sites suggest that the development of Burgess-like creatures stretches back into the early Cambrian.

Do the fossils of the Burgess Shale point to another breakthrough event in evolution? Stephen Jay Gould's 1989 book *Wonderful Life* tells the story of the Burgess Shale and makes a good case for a breakthrough. Traditionally, evolution has been seen as a gradual process that advances from more primitive and undifferentiated forms toward more precisely adapted forms, according to the survival of the fittest. However, far from being generic in their biological designs, the creatures of the Burgess Shale show rather subtle and complex adaptations to their ocean environment. They look to be much more "fine-tuned" than one would expect at the dawn of multicellular life. The Cambrian explosion accomplished a lot fast.

Moreover, the Burgess fauna represent a far greater diversity of fundamentally different biological designs than have appeared anytime since. To appreciate this point fully requires recognizing that biologists organize the world of living things into hierarchical taxonomy. At the top of this taxonomy sit the five great kingdoms of contemporary biology: animals, plants, fungi, complex single cells, and simple cells (including bacteria). The next division down consists of phyla. Although the exact number is somewhat debated, there are some twenty or thirty contemporary animal phyla, including such major phyla as annelids (worms and some related creatures), arthropods (insects, lobsters, crabs, and the like), mollusks (clams, snails, and so forth), and chordates (creatures with backbones and a few others).

The Burgess Shale discloses some fifteen to twenty animal organisms as different from today's creatures and one another as one would expect of separate phyla. In addition, the shale includes representatives of all the major modern animal phyla. Should taxonomists recognize each of these creatures as signifying a separate phylum? Certainly the structural contrast is there. This far-gone era hosted more fundamental structural diversity of animal forms than exists on earth today. Far from evolution branching out like a tree from some primal type, it is as though evolution conducted a "brainstorm," generating a diverse array and then pruning down.

With these points as encouragement, what about the five markers of a breakthrough event? Do they all make an appearance? Here is a tally:

Long search and limited progress. The search preceding the Cambrian explosion was the longest, least superficially promising search in the history of life: more than 3 billion years from the earliest signs of life to the beginning of the Cambrian era.

Precipitating event. Evolution passed through some bottleneck into the realm of complex multicellular organisms.

Rapid breakthrough. Once evolution went to work in this new arena, a plethora of life-forms quickly emerged. The Cambrian diversification occurred over a period of some 100 million years. This is long even on a geological time scale, to be sure, but it must be measured against the preceding 4 billion years on the one hand,

and the ensuing 500 million up to the present that added no major phyla.

Transformation. Many fundamentally different anatomical designs made their appearance in the Cambrian explosion, many indicative of phyla that do not exist today. Indeed, nature rarely produces something as radically distinct as a phylum. The generativity of the Cambrian explosion shows clearly in the diversity of phyla at the time, which included all the phyla extant today.

Thus the creatures of the Burgess Shale offer a second and larger-scale example of the earmarks of a breakthrough event. While of course these are chosen cases, there are many more found in the fossil record. As noted earlier, a number of paleontologists have argued that the evidence of the rocks denies the original notion of gradualist evolution. Rather, organisms appear to emerge in a "punctuate" manner, with long periods of little change for a class of organisms interrupted by brief periods of rapid extinction, diversification, shift in dominance, and so on. Such circumstances inherently have most of the features of a breakthrough event.

A final twist to the story of the Burgess Shale is that the obvious survivors were not always the actual survivors. Of two phyla of wormlike creatures, one thrives today in immense diversity, the other survives only in a few odd forms. Yet the first of these was much the rarer in the Burgess Shale. The same has happened many times in evolution. Remember, the dinosaurs held sway for more than 100 million years. They were the only sizable land animals in their day. As Wesson puts it, "there was no nondinosaur larger than a poodle during the dinosaur reign." Today, all gone. In the Burgess Shale, much of the future lay with some forms of life not at all dominant then. One of them was a rather innocuous creature called *Pikaia*, which had the precursor of a backbone, a so-called notochord. *Pikaia* could be the ancestor of all the chordates, including us.

14
Blind Minds and Smart Evolution

How People Outthink Evolution

Are people better at breakthrough thinking than evolution? At first thought, the answer might seem to be, "Yes!" After all, here we are amid jet planes, computers, democracies, symphonies, and even books about esoteric subjects like breakthrough thinking and evolution. Evolution never did *that*. At second thought, the answer might seem to be, "No!" Evolution did do "that"—it invented us. Moreover, the dazzling creativity of the biological world surrounds us.

At third thought, the comparison might simply seem hard to make, rather like comparing apples and oranges. The profligate creativity of plant and animal forms surely outstrips the sum of human insight. Evolution wins on quantity of diversity. On the other hand, human insight deals with a far broader range of topics than survival and reproduction. Human insight wins on diversity of diversity. A further part of the puzzle recognizes that evolution and human thought function on radically different time scales. Thinking generates possibilities faster than gene shuffling, but people have had a much shorter time to work with. What would be a fair comparison?

But these ways of puzzling over the problem focus on the track record and mask a better way of thinking about the problem—by focusing on how the human mind and evolution deal with Klondike challenges. We might even speak of their relative *Klondike intelligence,* meaning simply the degree to which they are smart about handling the four unreasonable Klondike traps of breakthrough thinking. By this criterion, the searches we human beings conduct, when we're working at our best, are much better tuned to cope with the hazards of Klondike spaces than the blind search process of Darwinian natural selection. Here are several ways in which minds are more insightful than genes.

Knowledge-driven thinking. Natural selection as usually conceived has no memory. The cards of genetic diversity are shuffled, dealt, and fall as they may. In contrast, people depend on their own individual memories and the cultural record as handed down to them personally or accessed through books, CD-ROMs, and other information resources. This means that people have a better chance of seeing what would otherwise be a Klondike problem as a homing problem. As emphasized in Part III, background knowledge often makes all the difference. Even when knowledge does not transform a Klondike problem entirely into a homing problem, it can push it in that direction by providing more clues to look for, more guideposts along the way to a solution.

Generating good bets. Relatedly, when humans engage in generate-and-test strategies, the items that get generated tend to be "good bets." Master chess players tend to generate quite good moves right away, before extensive search. Some research I myself conducted years ago shows that better poets tend to produce better candidate lines when they brainstorm options. From an information-processing standpoint, one can view this phenomenon as a reflection of when constraints get applied. A search-and-select process can generate candidates and then apply constraints to sort them into keepers and throwaways. Also, some constraints can get incorporated into the generating mechanism, so whatever candidates get generated already satisfy at least these constraints. With learning, this is what happens in human beings: We learn to get it closer to right the first time. In contrast, evolution as classically conceived simply shakes the genes up and takes a chance.

Brainstorming. When human inquirers face impasses, they often deliberately widen the scope of search, brainstorming very different approaches. Variation in natural selection is usually thought of as being blind to the pressures of the moment. It introduces the same degree of variation in easy times or hard.

Looking for generativity. Natural selection as usually conceived cares only about survival, not generativity. It does not seek out organisms with high potential for further diversification. In contrast, human inquirers often quite deliberately search for theories and frameworks with generative potential.

Pursuing promise. A basic limitation of natural selection is that it operates only through trying out organisms that can actually survive. In natural selection, there is no such thing as a promising first draft. Human creators commonly deal with promising first-draft inventions, theories, poems, and so on that are not at all viable as they stand. Natural selection would count them all as failures and look elsewhere.

Planning. Search in a one-story space of possibilities does not honor the richness of human cognition. Poets and inventors not only navigate in spaces of possible poems and inventions but think about what they are doing at a higher level—where they are going, how they might get there, what might go wrong. They think in possibility spaces of goals and plans as well as the "base space" of poems or inventions. Natural selection makes no allowance for such planning levels of search.

In these and other ways, human inquirers sometimes show more Klondike intelligence than natural selection does. That is, human searching is much more finely adapted to the challenges of Klondike spaces that work against achieving breakthrough events.

Maybe human beings evolved to be like that. Michael Ruse and Rupert Riedl among others argue that not only the morphology but also the epistemology of human beings can be seen as a product of Darwinian evolution. Ruse proposes that notions of causality, basic ideas about number, analogy making, and the notion that claims are verified by a convergence of evidence from various sources are all likely consequences of evolutionary processes, all very fundamental adaptive characteristics of mind with a genetic basis. To these elements of abstraction in thought, one might well add the power of self-management, including the management of searches.

But what environment is the mind adapted *to*? As angelfish move through a world of corals and foxes a world of brush and briar, the mind moves through a world of ideas, seeking ones that serve a mosaic of purposes. People think in terms not just of actualities but possibilities of varying payoff and promise. Much as angelfish and foxes have adapted through evolution to their physical environments, so very likely has the mind adapted to its mental one. In the spirit of Ruse and Riedl, human powers of abstraction and search management might in part be consequences of the genetic evolution of intelligence, selection specifically for characteristics that facilitate the search of Klondike spaces. By this measure, evolution has produced a mechanism far more sophisticated than itself in playing its own Klondike game.

Blind Minds

This is high praise for human thinking at its best. However, does human thought, art, science, and culture always advance through artful breakthrough thinking? Not at all. Sometimes, perhaps much of the time, matters of civilization play out more like the chaotic saga of life on earth.

As a "dessert" toward the end of his *The Selfish Gene*, zoologist Richard Dawkins developed the notion of *memes*, an analog of evolution not for biological forms but for the ideas of human beings. The philosopher Daniel Dennett picked up the theme, articulating it further in *Consciousness Explained*, and other authors have done likewise. The meme analogy follows the mechanism of evolution closely, but applies it to human mind and society. Thus Dawkins provides a language for talking in Darwinian terms about the social evolution of ideas.

Dawkins coined the word *meme* to mean "mental gene," a unit of thought that thrives in the ecology of human culture and communication. It could be an idea like democracy, an industrial practice like the smelting of iron, a medical practice like bleeding, a wise saying like "a stitch in time saves nine," or almost anything else. Dawkins's key point is that memes can evolve in much the same way that genes do, by breeding diverse offspring that in turn get selected for survival.

Imagine Benjamin Franklin making up homilies. One day he makes up three. The first says, "A penny saved is a penny earned." Another says, "A stitch in time saves seven." The third says, "Pay your debt to tomorrow." All three are memes—mental rather than genetic forms tossed into the ecology of human society. They will survive, or mutate and survive, or die out.

"Pay your debt to tomorrow" lacks poetry and compelling meaning. People forget it. "A penny saved is a penny earned" clicks. People remember it and pass it on. Like the shark and the cockroach, successful bioforms that have remained unaltered for millions of years, it hardly changes. "A stitch in time saves seven" totters along for a while. But at some point, someone makes up the variant, "A stitch in time saves nine." The greater euphony lets the mutant form propagate more widely, driving out the weaker competing "seven."

Presumably, such a mechanism underlies the social evolution of languages. From time to time, individuals spontaneously try out variations, typically with little thought. Some of the variations—contractions for example—catch people's fancy and get adopted and passed along. They in turn become the basis for other variations.

A key implication of Dawkins's meme notion is that the evolution of ideas does not depend on truth or excellence. It reflects survival power instead. The memes that thrive in the environment of mind and society are not necessarily the memes that make the most profound statements or express the richest artistic ideas. Disco music takes the day while fusion jazz only does okay. It is the same in the biological world. Insects are a far bigger success story than the more sophisticated and intelligent mammals. In similar vein, Stephen Jay Gould takes to task the popular way of describ-

ing evolution in terms of "the age of fishes, the age of reptiles, the age of mammals" and so on. Throughout the long
sweep of evolution, none of these creatures comes even
close to rivaling the population of the truly dominant life-
form on this planet. It's all the age of bacteria, Gould quips.
A personal favorite example of an obtuse meme but a
real survivor is the "right brain–left brain" notion.
Popularized a few years ago, this notion has spread into the
minds of almost everyone even idly concerned with psychology, creativity, or education. The general precept holds
that the right side of the brain manages the holistic, creative functions, while the left side takes responsibility for
the more analytic and linguistic functions. This is, of
course, a gross oversimplification and parody of the actual
findings. Moreover, it does not even make logical sense.
With such a division of labor, how do poets and mathematicians function creatively through language and nota-
tions? Nonetheless, the idea has enormous magnetism.
People love to believe it and draw questionable implications from it.

In keeping with Dawkins's analogy, memes have to handle
the four Klondike challenges. But how? As a blind process
like biological evolution, meme evolution does so in roughly
the same way, with a couple of interesting differences.

How memes deal with a wilderness of possibilities.
Memes breed by communication processes such as conversations, publishing, radio, and television, that pass them to
other minds. These processes often create variants through
distortion and mistakes people make in understanding. As
with genes, the search of possibilities done by memes
occurs in parallel, because each meme may produce
descendants in many minds, some variants surviving and
passing to other minds, some not.

In contrast to genetic evolution, things happen quickly. Teen slang comes and goes, and fashions of dress change in years and decades. One reason the pace is fast is that the host mind for a particular meme often tinkers with it, making it more viable. In the Benjamin Franklin fantasy, his original "A stitch in time saves seven" does not have to wait upon someone's slip of the tongue—the analog of a random mutation—to produce "nine." Some poetic soul might remake the homily.

How memes deal with clueless plateaus. As in genetic evolution, meme variants radiate from the ancestor in all directions. This fills plateaus with variants, and some on the edge of a plateau may spread to greener pastures.

How memes deal with narrow canyons of exploration. As with genetic evolution, memes evolve only through viable forms, that is, ideas that survive in one person's mind long enough to get passed to other minds in the original form or a variant. So far, memes appear to deal with the canyon problem no better than does evolution. However, as noted above, the host mind of a particular meme may tinker with it before passing it along, producing a revised meme that has hopped out of a narrow canyon. Thus, the minds that host memes introduce a dimension of flexibility to the process of meme evolution not present in the animal and plant bodies that host genes.

How memes deal with oases of false promise. Genetic evolution managed this through the spread of forms despite marginal viability. The same thing happens in meme evolution. For example, consider this variant of the Benjamin Franklin story. Suppose Franklin originally coins the phrase, "A stitch in time saves seven." Although not very catchy, it nonetheless slowly spreads to others, who pass it on intact or in tinkered versions—"saves six," "saves ten." Eventually someone produces "saves nine," which quickly spreads throughout society.

In summary, the evolution of memes offers a Darwinian mechanism of invention important in human culture. But with memes, unlike genes, the possessor of the meme can deliberately modify it. Accordingly, how much a pure process of meme variation and selection contributes depends on the likelihood of the hosts of a meme actively revising it. In the social evolution of languages and customs, intentional revision does not seem to play much of a role. However, occasionally someone will coin a word or establish a ritual that takes hold. In popular culture, memes thrive in relatively pure form because people do not reflect very much on what comes their way. They just reject it, or accept it and pass it along, perhaps with accidental variations.

But of course commercial efforts to capitalize on popular culture traffic in "designer memes," deliberate efforts to craft styles, genres, and so on, that will catch on and propagate widely. At the far extreme from a pure Darwinian meme process are intensively engineered technical inventions that get the calculated attention of their initial inventors and, later, of others who hope to produce their own superior and patentable variants. There is indeed a memelike spread of ideas from person to person, but with assiduous attention from the host minds.

Smart Evolution

If people are not always so smart, letting memes do part of the work of navigating Klondike spaces, perhaps evolution is not so dumb. Earlier, evolution was described in Darwinian terms of blind variation, selection, and inheritance, generating breakthroughs in Klondike space by brute force over millions of years. However, some evolutionists have challenged this picture and the classic mechanism of natural selection itself. They have argued either that more is going on than just natural selection, or at least that the story to be told around natural selection, involves elaborations beyond a roll of the genetic dice each generation. Evolution may be a process with more Klondike intelligence than appears on first account. It is worth reviewing some of the features of Klondike intelligence mentioned earlier for human beings, to see whether they might be found in evolution too.

Knowledge-driven thinking. In the human genome, up to 99 percent of the genes do not function to direct the development of the organism. They are passive, or latent. Similar figures apply to other organisms. It has been suggested that these passive genes constitute a repertoire of adaptations from the history of the organism that can be invoked. One phenomenon pointing to this is the rapid response many insect species make to insecticides. Some develop inheritable resistance within a single harvest season. This suggests the activation of existing latent resources more than a haphazard genetic search for a new solution. In the language of the four operations of breakthrough thinking, it's a matter of detection—detecting which latent resource fits the circumstances.

Generating good bets. Human thinkers generate likely candidates for the challenges they face. In laboratory studies, bacteria often adapt in very few generations to use, as food, substances they normally cannot use at all. This suggests that whatever trial variants are generated are "good bets" to take advantage of the available nutrients. The "good bets" may come from a latent repertoire, as mentioned in the previous paragraph.

Brainstorming. Human thinkers at an impasse sometimes generate diverse possibilities. Under a variety of stress conditions, many organisms can increase the rate of genetic variation, "brainstorming" as it were, to find a new path. Plants often are genetically altered in response to environmental changes. For example, environmental pressures, especially temperature, produce heritable variations in flax. Features that change include size, weight, features of the seed capsule, certain chemical products, and amount of nucleic acid. Thus, under some circumstances, evolution increases its roving to explore more options.

Looking for generativity. Human inquirers often seek out specifically generative theories and frameworks, ones that lead on to other things. In the biological world, high variation is especially advantageous in times of rapid change, with consequent adaptation pressure. In conditions that call for diversification, organisms apparently can become genetically predisposed to vary genetically: Change itself becomes an agenda. Experiments in selecting bacteria for variability show that the bacteria increased in variability, preserving variability itself as a trait. Another organism of special interest is cichlid fish, a group of freshwater species found in Africa, Mexico, and Hawaii. In one species, the young of a single brood show major variations in teeth and in the structure of the digestive system. Perhaps because of this high variability in the cichlid fish generally, over a period of a million years they have differentiated into some 300 species, many with very different lifestyles, in seven East African lakes.

Pursuing promise. Human inquirers often try out imagined versions or prototypes before producing the real thing. Something structurally analogous can happen in evolution. Characteristics an organism acquires during its lifetime, through learning, muscular strengthening, or other means, can provide the organism with a rough adaptation to a niche. Imagine, for example, that birds of a particular species have discovered an ample source of food in the insects lurking between the rough bark of trees and have begun to feed there. This rough adaptation is structurally analogous to the human's pursuit of promise.

An organism within the scope of its lifetime can try out a new niche—indeed several—and often withdraw unscathed if unsuccessful. If the organism proves somewhat successful, it may occupy the niche, and its progeny and others of its kind follow suit. Then adaptive pressures go to work, over millennia making the acquired adaptations genetic and adding others. The birds who discovered the insects might evolve into modern-day woodpeckers, anatomically and behaviorally specialized for such feeding. This is called the Baldwin effect after James Mark Baldwin, a philosopher and psychologist of the late nineteenth and early twentieth century at Princeton.

Of course, the features of evolution mentioned earlier do not always challenge natural selection as the basic underlying mechanism. However, they at least point out that natural selection operates within a supportive context: a repertoire of passive genes, creatures that can learn and hence try out new niches in their lifetimes, and so on. Thus, to understand how the mechanism of evolution searches, it is simply not enough to speak of blind variation, selective survival and reproduction, and inheritance, even if these form the core of the process.

In summary, evolution after all has a fair measure of

Klondike intelligence. There is no mystical or teleological intent to such a claim. No master planner behind evolution is being proposed. Rather, evolution is smart in the sense defined earlier: The process of evolution in a number of ways is well adapted to searching in a Klondike space, much better adapted than the classic Darwinian trial-and-error process would suggest.

Such adaptations are, of course, carried by genes. The long history of life on earth has led genomes to adapt not just to various environments but to the contingencies of evolution itself. Genes are not dumb, they are smart.

15

Why We're Stuck with Breakthrough Thinking

Bad Luck and Beyond

What a nuisance! Both human thought and the process of evolution seem to get stuck all too often with unreasonable problems that call for breakthroughs. All too often, people have to think in a different and less comfortable way. The mechanistic process of evolution does not, of course, suffer headaches the way we do, but it does have to cope with the bother of it all.

So why this nuisance? Why is it that maneuvering in a Klondike space to accomplish a breakthrough so often comes up?

Reason 1. The odds against order. Partly, it's inevitable bad luck. Chance plays a role and an important one. A wishful thinker might ask for a neater world where homing spaces were the rule and Klondike spaces exotic rarities like white tigers. However, this is not a reasonable expectation. Recall what a simple structure a homing space is. The ideal homing space presents a landscape rich with clues that point clearly to the one pinnacle to be attained. The problem solver astute enough to read the clues can progress steadily up the hill to the prize. Real-world problems, random rabbits plucked out of whatever hat the world presents us with, are bound on the average to bring with them irregular possibility spaces with a range of troublesome Klondike characteristics.

But it's not just inevitable bad luck. If we look into the way problems requiring breakthrough thinking come up, we discover that they arise not just by chance but because of our very success in solving the homing problems. Breakthrough challenges are the legacy of homing successes. To see how this happens, it's important to understand a very general concept from systems theory called self-organizing criticality.

Self-Organizing Criticality

For the same reason that tires squeal, Klondike problems proliferate in human invention. For the same reason that earthquakes threaten California, Klondike fitness landscapes appear in biological evolution. Klondike possibility spaces do not just happen to be there. Something about the circumstances generates their inconveniences.

Self-organizing criticality is a phenomenon that has received much attention in contemporary studies of complex systems. The standard example involves a pile of sand. Imagine trickling sand down slowly to create a pile, which quickly assumes a conical shape. As the sand descends, an interesting pattern of stasis and change develops. A few grains generally do not affect the pile very much. They lodge at various points near the peak. However, as the peak comes to carry more and more grains, eventually something has to give way. A landslide occurs that relieves the burden. As the trickle of sand continues, this pattern of stasis and change repeats, with smaller and occasionally larger landslides.

The sand pile is a paradigm example of *self-organizing criticality*. The criticality part refers to the punctuate nature of the change. The sudden landslides are critical events that interrupt the stasis. The self-organizing part refers to the way that the whole process converges on such critical events. As the sand trickles down, the physical structure of the pile inherently draws closer and closer to a critical event.

Tires squeal and chalk squeaks on blackboards for the same reason. As sheer force builds up between one surface and another, at first the surfaces stick, but they cannot stick forever, so eventually they suddenly slip against one another. Once they slip, relieving the sheer force, they stick again for a while. What happens on a microscale with tires and chalk happens on a macroscale with earthquakes. Tectonic plates in slow but inexorable motion press against one another, and the pressure mounts. Eventually, an earthquake occurs, a critical event that relieves the pressure, and the buildup begins again.

Why the World Demands Breakthroughs

Besides squealing tires and earthquakes, self-organizing criticality has another daughter. Klondike landscapes in the worlds of biological evolution and human inquiry are in part the consequence of self-organizing criticality. This seems to happen through at least two processes, each in its own way analogous to the pattern of buildup and sudden release produced by trickling sand onto a sand pile. So besides the basic *odds against order* reason for unreasonable problems, there are two more.

Reason 2. Easy answers first. Any adaptive situation full of problems to be addressed has a spectrum of more approachable and less approachable problems. Among the less approachable ones are the Klondike problems. The more approachable problems are dealt with first, leaving a Klondike mountain range demanding attention. For instance, surgery on limbs and many organs developed early, because they are not critical to the moment-to-moment functioning of the body. But heart surgery developed late, because the heart had to keep pumping, or have something else like a heart-lung machine do its job, during the very operation.

The process of easy answers first does not create the Klondike landscapes, but it systematically leads the problem-solving agent toward the Klondike portions of the landscape as more tractable portions get settled, building up toward critical events.

Reason 3. Grooves entail walls. In any adaptive situation, further adjustments and refinements follow upon initial adaptation. Artists settle into their styles. Physicists develop repertoires of techniques and backlogs of questions that reflect their specializations. In the biological world, organisms develop nuanced adaptations to the peculiarities of their ecological niches. They are all creating grooves—and grooves not to be despised, but very effective ones that function as relatively safe and generative venues.

However, grooves entail walls. The adaptations themselves amount to partial commitments to their settings. They are enabling within their scope but limiting beyond their scope. They create oases and canyons, hard to depart from if there is a need to do so.

Because of *odds against order, easy answers first,* and *grooves entail walls,* the human endeavor to find and solve problems tends to lead us to Klondike situations and create more Klondike situations. Similar things happen in the biological world, without any implication of a guiding vision. The phenomenon of self-organizing criticality from earthquakes to Einstein makes it plain that the pattern of sudden discovery characteristic of breakthrough events is one member of a much larger class of punctuate phenomena.

What is special about breakthrough events in the human and biological world is that chalk does not squeak, nor earthquakes shake, in order to accomplish anything. Wildernesses and plateaus, canyons and oases mean nothing to chalk or tectonic plates. Squeaks and earthquakes just happen to happen the way they do. In contrast, humans aim to invent better and more varied ways of doing and expressing things, sometimes blindly but often with great intentionality and effort. Evolution is itself adapted to develop and refine ways and forms of survival, blind process though it is.

Without Newton and his followers constructing the comfortable and largely effective groove of classical physics, there could not have been an Einstein. Without the classic discipline of the French Academy, there could not have been the revolution of Impressionism. Even if there were something that looked like Impressionism, it would not carry the meaning and impact that Impressionism did and does. The Klondike realities that human inquirers must face are in large part self-created on individual and social scales—not through some flaw in human nature, some original cognitive sin, but as an inherent and inevitable part of the dynamics of problem finding and problem solving.

Notes

1. Thinking like Leonardo

Thinking Takes Flight

Leonardo da Vinci's helicopter design: Edward MacCurdy, *The Mind of Leonardo da Vinci* (London: Jonathan Cape, 1928). First quote from Leonardo da Vinci: same source, p. 268. Second quote, p. 275.

The Wright brothers' development of the propeller: Tom Crouch, *The Bishop's Boys: A Life of the Wright Brothers* (New York: Norton, 1989), Chap. 18.

Eureka!

Archimedes' eureka experience: Arthur Koestler, *The Act of Creation* (New York: Dell, 1964).

Malthus Kicks a Field Goal Twice

Darwin's account of the discovery of the principle of natural selection: Charles Darwin, *The Life and Letters of Charles Darwin*, vol. 1, Francis Darwin, ed. (New York: Appleton, 1888), p. 68.

Darwin's not recognizing the full scope of his discovery at first: Howard Gruber, *Darwin on Man: A Psychological Study of Scientific Creativity* (New York: Dutton, 1974).

Alfred Russel Wallace's account of his discovery of the principle of natural selection: Alfred Russel Wallace, *My Life*, vol. 1 (New York: Dodd, Mead, 1905), pp. 361–362. The emphasis is Wallace's.

Ninety-Nine Percent Perspiration

Edison's search for a filament: Neil Baldwin, *Edison: Inventing the Century*, (New York: Hyperion, 1995).

Robert Friedel and Paul Israel, *Edison's Electric Light* (New Brunswick, N.J.: Rutgers University Press, 1986).

W. B. Carlson and M. Gorman, "A Cognitive Framework to Understand Technological Creativity: Bell, Edison, and the Telephone," in R. J. Weber and D. N. Perkins, eds., *Inventive Minds: Creativity in Technology* (New York: Oxford University Press, 1992), pp. 48–79.

How Stoneflies Came to Fly

The story of the stonefly: James H. Marden "How Insects Learned to Fly," *The Sciences*, 35, no. 6 (1995): 26–30.

Mother Nature Breaks Through

Punctuated equilibrium: Stephen J. Gould, *The Panda's Thumb: More Reflections in Natural History* (New York: Norton, 1980), chaps. 17–18.

The Mainspring Question

The quote from Plato about inspiration: *Ion*, 534, in Edith Hamilton and Huntington Cairns, eds., Lane Cooper (trans.), *The Collected Dialogues of Plato* (Princeton, N.J.: Princeton University Press, 1961).

Whether and how intelligence can be improved: David Perkins, *Outsmarting IQ: The Emerging Science of Learnable Intelligence* (New York: Free Press, 1995).

2. From Sufi Tales to James Bond Thrillers

How to Be Blind

Sufi teaching tales: Idries Shah, *Tales of the Dervishes* (New York: E. P. Dutton, 1970), p. 20.

Be Reasonable!

The source of the three-letter word puzzle: Pierre Berloquin, *100 Games of Logic* (New York: Barnes & Noble, 1977), p. 14, game 11.

The quote from Margaret Boden: *The Creative Mind: Myths and Mechanisms* (New York: Basic Books, 1991), p. 40.

Cryptarithmetic problems: See Alan Newell and Herbert Simon, *Human Problem Solving* (Englewood Cliffs, N.J.: Prentice-Hall, 1972).

The Hunting of the Snap

Findings in support of insight problems involving a cognitive snap: Janet Metcalfe, "Premonitions of Insight Predict Impending Error," *J Exp. Psychology: Learning Memory and Cognition*, 12, (1986): 623–634.

Janet Metcalfe and David Wiebe, "Intuition in Insight and Noninsight Problem Solving." *Memory & Cognition*, 15, (1987): 238–246.

Janet Davidson, "The Suddenness of Insight," in Robert Sternberg and Janet E. Davidson, eds., *The Nature of Insight* (Cambridge, Mass.: MIT Press, 1995), pp. 125–155.

Robert S. Lockhart, Mary Lamon, and Mary Gick, "Conceptual Transfer in Simple Insight Problems," *Memory & Cognition*, 16, (1988): 36–44.

Insight on a Platter
Marvin Minsky on humor: Marvin Minsky, "Jokes and the Logic of the Cognitive Unconscious," in R. Groner, M. Groner, and W. F. Bischof, eds., *Methods of Heuristics* (Hillsdale, N.J.: Lawrence Erlbaum Associates, 1983), pp. 171–194.

3. The Logic of Lucking Out

Gutenberg Lucks Out
Gutenberg's invention of movable type and the printing press: Koestler, *Act of Creation* (cited earlier in Chapter 1).

The Luck of the Klondike
The view of breakthrough thinking presented here and throughout the rest of the book has been developed by me through a series of technical articles over the years. Key articles include:

David N. Perkins, "The topography of invention," in R. J. Weber and D. N. Perkins, eds., *Inventive Minds: Creativity in Technology* (New York: Oxford University Press, 1992), pp. 238–250.

David Perkins, "Creativity: Beyond the Darwinian Paradigm," in M. Boden, ed., *Dimensions of Creativity* (Cambridge, Mass.: MIT Press, 1994), pp. 119–142.

David Perkins, "Insights in Minds and Genes," in Robert J. Sternberg and Jane E. Davidson, eds., *The Nature of Insight* (Cambridge, Mass.: MIT Press, 1995), pp. 495–533.

David Perkins, "The Evolution of Adaptive Form," in John Ziman, *Technological Innovation as an Evolutionary Process* (Cambridge, England: Cambridge University Press, forthcoming).

Six Paper Klondikes
The Two Strings problem is one of several insight problems studied by an early investigator of the genre: Norman R. Maier, *Problem Solving and Creativity in Individuals and Groups* (Belmont, Calif.: Brooks/Cole, 1970).

The Extended Family is adapted from Berloquin, *100 Games of Logic* (cited earlier for Chapter 2).

4. Is There a Science of Breakthrough Thinking?

Everything Is Like a Game of Chess
Problem solving as a search through a space of possibilities: Newell and Simon, *Human Problem Solving* (cited earlier for Chapter 2).

Jack London's Klondike
Jack London's short story "All Gold Canyon" can be found in Howard Lachtman, ed., *Young Wolf: The Early Adventure Stories of Jack London* (Santa Barbara, Calif.: Capra Press, 1984).

5. Thinking's Big Bang

Roving Far and Wide
The development of the zeolite catalyst: Edward Rosinski, "The Origin and Development of the First Zeolite Catalyst for Petroleum Cracking," in R. J. Weber and D. N. Perkins, eds., *Inventive Minds: Creativity in Technology* (New York: Oxford University Press, 1992), pp. 166–177.
The development of ivermectin: William Campbell (1992), "The Genesis of the Antiparasitic Drug Ivermectin," in R. J. Weber and D. N. Perkins, eds., *Inventive Minds: Creativity in Technology* (New York: Oxford University Press, 1992), pp. 194–214.
Combinatorial chemistry: Matthew J. Plunkett and Jonathan A. Ellman, "Combinatorial Chemistry and New Drugs," *Scientific American*, 276, no. 4 (1997): 68–73.

How to Court Lady Luck
The conference on invention: Robert Weber and David Perkins, eds., *Inventive Minds: Creativity in Technology* (New York: Oxford University Press, 1992).
For a full discussion of the general characteristics of these inventors and their process, see Perkins and Weber, "Effable Invention," in Weber and Perkins, eds., *Inventive Minds*, (cited above), pp. 317–336.

The Breakthrough Logic of Brainstorming
Brainstorming: Alex Osborn, *Applied Imagination* (New York: Charles Scribner's Sons, 1953).
Robert Sutton and Andrew Hargadon, "Brainstorming Groups in Context: Effectiveness in a Product Design Firm," *Administrative Quarterly*, 41, no. 4 (1996): 685–718.

Theme and Variations
Wallace, Stevens "Thirteen Ways of Looking at a Blackbird," in *The Collected Poems of Wallace Stevens* (New York: Knopf, 1957).

"Ars Poetica": in Archibald MacLeish, *Collected Poems of Archibald MacLeish* (Boston: Houghton Mifflin, 1962).

The work of Hieronymus Bosch: Benedikt Taschen, *Bosch* (Cologne, Germany: Druckhaus Cramer, 1994).

One Hundred Views of Mount Fuji: see Henry Smith II, *Hokusai: One Hundred Views of Mount Fuji* (New York: George Braziller, 1988).

The Breakthrough Logic of Variations
"On the Vanity of Earthly Greatness": Arthur Guiterman, *Gaily the Troubadour* (New York: Dutton, 1936).

The Art of the Arbitrary
A number of techniques for provoking ideas, including random input and stepping-stones, are discussed in Edward de Bono, *Lateral Thinking: Creativity Step by Step* (New York: Harper & Row, 1970).

6. Clues for the Clueless

The Dog Did Nothing in the Nighttime
The Sherlock Holmes story "Silver Blaze": Arthur Conan Doyle, "Silver Blaze," in *The Complete Sherlock Holmes* (Garden City, N.Y.: Doubleday, 1960).

Einstein as Sherlock Holmes
Einstein's *Gedanken* experiment: A. I. Miller, *Imagery in Scientific Thought: Creating 20th-Century Physics* (Boston: Birkhauser, 1984).

Gerald Holton, "On Trying to Understand Scientific Genius," *Thematic Origins of Scientific Thought: Kepler to Einstein* (Cambridge, Mass.: Harvard University Press, 1973), chap. 10.

The quote from Einstein is from his *Autobiographical Notes* (1949), as quoted in Gerald Holton, *Thematic Origins of Scientific Thought: Kepler to Einstein,* revised ed. (Cambridge, Mass.: Harvard University Press, 1988), p. 31.

When Computers Think Better than People
Artificial intelligence and a classic theorem of geometry: Boden, *Creative Mind* (cited earlier in Chapter 2), pp. 104–110.

When Imagination Signifies Nothing
Synectics: William J. J. Gordon, *Synectics: The Development of Creative Capacity* (New York: Harper & Row, 1961).

The discussion of the drag alarm problem and shortcomings of analogical reasoning: David Perkins "Novel Remote Analogies Seldom Contribute to Discovery," *Journal of Creative Behavior,* 17, (1983): 223–239.

Creative Cluelessness
Preinventive forms and the Geneplore model: Ronald Finke, "Creative Insight and Preinventive Forms," in Robert J. Sternberg and Janet E. Davidson, eds., *The Nature of Insight* (Cambridge, Mass.: MIT Press, 1995).

Ronald Finke, Thomas Ward, and Steven Smith, *Creative Cognition: Theory, Research and Applications* (Cambridge, Mass.: MIT Press, 1992).

7. Walking through Walls

Einstein on the MTA
Einstein's first relativity paper, "On the Electrodynamics of Moving Bodies": Albert Einstein, "Zur Elektrodynamik bewegter Korper," *Annalen der Physik*, 17, (1905): 891–921.

Reframing Problems
The work of de Bono: See, for instance, his *Lateral Thinking* (cited earlier for Chapter 5); *Six Thinking Hats* (New York: Viking, 1986); and *The Mechanism of the Mind* (London: Penguin, 1990).

Conditions that cue the reframing of a problem: Stellan Ohlsson, "Restructuring Revisited: An Information Processing Theory of Restructuring and Insight," *Scandinavian Journal of Psychology*, 25 (1984): 117–129.

The use of analogies in molecular biology research: Kevin Dunbar, "How Scientists Really Reason: Scientific Reasoning in Real World Laboratories," in Robert J. Sternberg and Janet. E. Davidson, eds., *The Nature of Insight* (Cambridge, Mass.: MIT Press, 1995), pp. 365–395.

Three Problems of Reframing
The Giraffes and Ostriches problem is drawn from Davidson, "Suddenness of Insight," (cited earlier for Chapter 2).

Solutions by Reframing
The mind's eye: S. M. Kosslyn, W. L. Thompson, I. J. Kim, and N. M. Alpert, "Topographical Representations of Mental Images in Primary Visual Cortex" *Nature*, 378 (1995): 496–498.

The discussion of The Four Beetles problem: David Perkins, *Knowledge as Design* (Hillsdale, N.J.: Lawrence Erlbaum Associates, 1986), chap. 7.

The Art and Craft of Reframing
Planning spaces are discussed in Newell and Simon, *Human Problem Solving* (cited earlier for Chapter 2).

8. Anything but That

Thinking Under Pressure
The story of the student and the barometer is recounted by Alexander Calandra, Teacher's Edition of *Current Science*, 49, no. 14 (1964). I encountered it in Murray Gell-Mann, *The Quark and the Jaguar: Adventures in the Simple and the Complex* (New York: Freeman, 1994), pp. 270–273. The solutions are direct quotes.
Convergent insight versus divergent insight: Finke, "Creative Insight and Preinventive Forms," (cited earlier for Chapter 6).

Decentering, Fourth Century B.C.E.
Einstein on formulating problems: Albert Einstein and L. Infeld, *The Evolution of Physics* (New York: Simon & Schuster, 1938), p. 92.

Decentering in the Modern Era
The reentry problem is discussed in: Michael Collins, *Liftoff: The Story of America's Adventure in Space* (New York: Grove Press, 1988).
Repurposing and the development of Post-its is discussed in Robert J. Weber, *Forks, Phonographs, and Hot Air Balloons: A Field Guide to Inventive Thinking* (New York: Oxford University Press, 1992).

The Importance of Problem Finding
On problem finding in general, see Eileen Jay and David Perkins, "Creativity's Compass: A Review of Problem Finding," in M. Runco, ed., *The Creativity Research Handbook*, vol. 1 (Cresskill, N.J: Hampton Press, 1997), pp. 257–293.
Studies of art students and problem finding: Jacob Getzels and Mihaly Csikszentmihalyi, *The Creative Vision: A Longitudinal Study of Problem Finding in Art* (New York: Wiley, 1976).
The follow-up study of professional creativity: Jacob Getzels and Mihaly Csikszentmihalyi, "Creativity and Problem Finding in Art," in F. H. Farley and R. W. Neperud, eds., *The Foundations of Aesthetics, Art, and Art Education* (New York: Praeger, 1988), pp. 91–116.
Writing and problem finding: Michael T. Moore, "The Relationship Between the Originality of Essays and Variables in the Problem-Discovery Process: A Study of Creative and Non-Creative Middle School Students," *Research in Teaching of English*, 19, no. 1 (1985): 84–95.
The importance of a sense of promisingness: Carl Bereiter and Marlene Scardamalia, *Surpassing Ourselves: An Inquiry into the Nature and Implications of Expertise* (Chicago: Open Court, 1993).

Opportunistic Invention
Creativity in dolphins: Karen Pryor, Richard Haag, and Joe O'Reilly, "Dolphin Cognition and Behavior: A Comparative Approach," in *On Behavior* (North Bend, Wash.: Sunshine Books, 1995).

9. Breakthroughs on Trial

Witnesses for the Prosecution
Discovery through reasoning: Pat Langley, Herbert Simon, Gary Bradshaw, and Jan Zytkow, *Scientific Discovery: Computational Explorations of the Creative Processes* (Cambridge, Mass.: MIT Press, 1987).

Kepler as Computer
The data table is from Langley et al., *Scientific Discovery* (cited above), p. 69.

Max Planck in a Reasonable Mood
Max Planck and blackbody radiation: Langley et al., *Scientific Discovery*, p. 47 and on.

Lavoisier's Leap
Lavoisier's challenge to phlogiston: Langley et al., *Scientific Discovery*, pp. 223–255; quotes regarding Lavoisier's reasoning are on pp. 250 and 251.

10. Is There a Mental Overdrive?

The Mysterious Magical Cognitive Snap

Bisociation Good and Bad
Bisociation: Koestler, *Act of Creation* (cited earlier for Chapter 1).
Normal science and paradigm changes: Thomas Kuhn, *The Structure of Scientific Revolutions* (Chicago: University of Chicago Press, 1962).

A Selective Process Theory of Insight
The selective process theory of insight: Davidson, (cited earlier for Chapter 2).
The triarchic theory of intelligence: Robert Sternberg, *Beyond IQ: A Triarchic Theory of Human Intelligence* (New York: Cambridge University Press, 1985).
Problem sources: The Checker Players is from Robert Sternberg and Janet Davidson, "The Mind of the Puzzler." *Psychology Today*, 16 (1982): 37–44.

Three Steaks is from Janet Davidson and Robert Sternberg "What is Insight?" *Educational Horizons,* 64, (1986): 177–179.

The Garment Rack is from Norman Maier "Reasoning in Humans III: The Mechanisms of Equivalent Stimuli and of Reasoning," *Journal of Experimental Psychology,* 35 (1945): 349–360.

These problems are also discussed in Davidson, "Suddenness of Insight."

What Selective Process Proves and What It Doesn't

The quote from Janet Davidson: Davidson, "Suddenness of Insight," p. 133.

Findings from Davidson and Sternberg's investigations of the three-process theory: Davidson, "Suddenness of Insight."

Incubation: The Pause That Refreshes

Studies of incubation: Robert Olton and D. M. Johnson, "Mechanisms of Incubation in Creative Problem Solving," *American Journal of Psychology,* 89, no. 4 (1976): 617–630.

Robert Olton, "Experimental Studies of Incubation: Searching for the Elusive," *Journal of Creative Behavior,* 13, no. 1 (1979): 9–22.

11. Presence of Mind

The Challenge of Inert Knowledge

Experiments on inert knowledge: J. D. Bransford, J. Franks, N. J. Vye, and R. D. Sherwood, "New Approaches to Instruction: Because Wisdom Can't Be Told," in Stella Vosniadou and Andrew Ortony, eds., *Similarity and Analogical Reasoning* (pp. 470–497). (New York: Cambridge University Press, 1989).

G. A. Perfetto, J. D. Bransford, and J. J. Franks, "Constraints on Access in a Problem Solving Context," *Memory & Cognition,* 11, no. 1 (1983): 24–31. The Polygamist puzzle is also from this article.

What Alexander Fleming Already Knew

The discovery of penicillin: Weber, *Forks, Phonographs and Hot Air Balloons* (cited earlier for Chapter 8).

And more on Fleming's discovery of penicillin: *Encyclopedia Americana, International Edition* (Danbury, Conn.: Grolier, 1997), Grolier, vol 11, p. 388.

How Pattern Priming Works

Quote from Pasteur: John Bartlett, *Familiar Quotations,* 16th ed. (Boston: Little, Brown, 1982), p. 502; originally quoted in Rene Valley-Radot, *The Life of Pasteur* (1927).

The Zeigarnik effect: Bluma Zeigarnik, "Uber das Behalten von Erledigten und Unerledigten Handlungen," *Psychologisches Forschung*, 9 (1927): 1–85.

Further research on the Zeigarnik effect: Colleen M. Seifert, David E. Meyer, Natalie Davidson, Andrea Patalano, and Ilan Yaniv, "Demystification of Cognitive Insight: Opportunistic Assimilation and the Prepared-Mind Perspective," in Robert J. Sternberg and Janet E. Davidson, eds., *The Nature of Insight* (Cambridge, Mass.: MIT Press, 1995), pp. 65–124.

Is Pattern Priming Effective?
Further experiments by Seifert and colleagues are described in Seifert et al., "Demystification of Cognitive Insight" (cited above).

Knowledge of the Genre of Insight Puzzles
These five puzzles are drawn from an article by Robert W. Weisberg, "Prolegomena to Theories of Insight in Problems Solving: A Taxonomy of Problems," in Robert J. Sternberg and Janet E. Davidson, eds., *The Nature of Insight* (Cambridge, Mass.: MIT Press, 1995), pp. 157–196.

12. Knowing Too Much and Forgetting Enough

Knowing Too Much for Your Own Good
A Disposable Spill-Proof Coffee Cup task: Steven Smith, "Getting into and Out of Mental Ruts: A Theory of Fixation, Incubation, and Insight," in Robert J. Sternberg and Janet E. Davidson, eds., *The Nature of Insight* (Cambridge, Mass.: MIT Press, 1995), p. 229–251.

Fertile Forgetting
The study of word rebuses: Smith, "Getting into and out of Mental Ruts" (cited above).

Unforgettable
The study of drawing alien monsters: Finke et al., *Creative Cognition* (cited earlier for Chapter 6).

A Klondike Psychology Playground
Puzzle sources: The Marriage Law is adapted from Raymond M. Smullyan, *What Is the Name of This Book? The Riddle of Dracula and Other Logic Puzzles* (Englewood Cliffs, N.J.: Prentice-Hall, 1978).

The Apple Orchard is adapted from Weisberg, "A Taxonomy of Problems" (cited earlier for Chapter 12).

Household Tasks is adapted from Davidson, "Suddenness of Insight" (cited earlier for Chapter 2).

13. Evolution Breaks Through

Ten Million Year Snaps
The Klondike view of evolutionary breakthroughs presented in this and subsequent chapters has been developed by me over several years through technical articles: Perkins, "Creativity: Beyond the Darwinian Paradigm" (cited earlier for Chapter 3).

Perkins, "Insight in Minds and Genes" (cited earlier for Chapter 3).

Perkins, "The Evolution of Adaptive Form" (cited earlier for Chapter 3).

The Breakthrough of Birds
Possible functionality of feathers: Gould, *Panda's Thumb* (cited earlier for Chapter 1).

Robert Wesson, *Beyond Natural Selection* (Cambridge, Mass.: MIT Press, 1991), p. 47.

Feathers appearing abruptly in the fossil record: Wesson *Beyond Natural Selection* (cited above).

Ground birds assuming the role of major predators in South America: Stephen Jay Gould, *Wonderful Life: The Burgess Shale and the Nature of History* (New York: Norton, 1989), pp. 297–298.

The Breakthrough of the Burgess Shale
The story of the Burgess Shale: Gould, *Wonderful Life* (cited above). In particular, reasons for the good preservation of organisms in the Burgess shale are from pp. 224–227; the diversity of Burgess Shale organisms and the resulting taxonomic challenges are drawn from pp. 99, 208–209; and the worm with a notochord in the Burgess Shale is discussed on pp. 292–295.

The quote "There was no nondinosaur . . ." comes from Wesson, *Beyond Natural Selection* (cited above), p. 214.

14. Blind Minds and Smart Evolution

How People Outthink Evolution
On master chess players generating good moves: Neil Charness, "Expertise in Chess: The Balance Between Knowledge and Search," in K. A. Ericsson and J. Smith, eds., *Toward a General Theory of Expertise: Prospects and Limits* (New York: Cambridge University Press, 1991), pp. 39–63.

Experienced poets generating richer material: David Perkins, *The Mind's Best Work* (Cambridge, Mass.: Harvard University Press, 1981).

Human epistemology has evolved too: Michael Ruse, *Taking Darwin Seriously: A Naturalistic Approach to Philosophy* (Oxford: Blackwell, 1986).

Riedl, R., *Biology of Knowledge: The Evolutionary Basis of Reason* (New York: Wiley, 1984).

Blind Minds

On memes: Richard Dawkins, *The Selfish Gene* (New York: Oxford University Press, 1976).

Daniel Dennett, *Consciousness Explained* (Boston: Little, Brown, 1991).

Aaron Lynch, *Thought Contagion: How Beliefs Spread Through Society* (New York: Basic Books, 1999).

The "age of fishes . . ." comment is by Stephen Jay Gould from his keynote presentation to the Sixth International Conference on Thinking, Cambridge, Mass., July 17–22, 1994.

On the vacuous left brain-right brain notion: Howard Gardner, *The Shattered Mind* (New York: Knopf 1975).

Perkins, *Mind's Best Work* (cited above).

Smart Evolution

Several mechanisms have been identified that allow for inheritance of adaptations without genetic modification, although genetic change remains the mainstay. See Eva Jablonka and Marion J. Lamb, *Epigenetic Inheritance and Evolution: The Lamarckian Dimension* (New York: Oxford University Press, 1995).

The points in this section are, by and large, drawn from Wesson, *Beyond Natural Selection* (cited earlier for Chapter 13). In particular: Passive, or latent genes: p. 229. Rapid adaptation of insect populations to insecticides: pp. 236–237. Quick adaptation by bacteria to using previously unassimilable food substances: p. 234. Increase in the rate of genetic variation under stress conditions: pp. 238–239. Selection of bacteria for variability as a trait: p. 211. High variability of cichlid fish: p. 211. Exploration and adaptation through learning during one's lifetime leading to occupying niches and evolving into them, p. 241.

15. Why We're Stuck with Breakthrough Thinking

Self-Organizing Criticality

Summaries of the idea of self-organizing criticality in: M. Mitchell Waldrop, *The Emerging Science at the Edge of Order and Chaos* (New York: Simon & Schuster, 1992).

Gell-Mann, *The Quark and the Jaguar* (cited earlier for Chapter 8).

Permissions _____

Leonardo da Vinci's helicopter design: Edward MacCurdy, *The Mind of Leonardo da Vinci*. London: Jonathan Cape, 1928. Reprinted with permission of Bibliotheque de L'Institut de France, Paris.

The work of Heironymous Bosch: Benedikt Taschen, *Bosch*. Germany: Druckhaus Cramer Gmbh, 1994. Studies of Monsters reprinted with permission of the Ashmoleum Museum, Oxford. Tree Man Sketch reprinted with permission of Albertina Museum, Vienna.

One Hundred Views of Mount Fuji: see Henry Smith II, *Hokusai: One Hundred Views of Mount Fuji*. New York: George Braziller, 1988. Reprinted with permission of David N. Perkins.

"On the Vanity of Earthly Greatness": Arthur Guiterman, *Gaily the Troubadour*. New York: E. P. Dutton, 1936.

Data table from Pat Langley, Herbert Simon, Gary Bradshaw, and Jan Zytkow, *Scientific Discovery: Computational Explorations of the Creative Processes*. Cambridge, Mass.: MIT Press, 1989. Reprinted with permission of MIT Press.

Disposable spillproof coffee cup example shown to subjects to induce fixation in Robert J. Sternberg and Janet E. Davidson, eds., *The Nature of Insight*, pp. 125–155. Cambridge, Mass.: MIT Press, 1995. Reprinted with permission of MIT Press.

Set of object parts in experiments on creative invention in Ronald Finke, Thomas Ward, and Steven Smith, *Creative Cognition: Theory, Research and Applications*. Cambridge, Mass.: MIT Press, 1992. Reprinted with permission of MIT Press.

The Contact Lens Remover in Ronald Finke, Thomas Ward, and Steven Smith, *Creative Cognition: Theory, Research and Applications.* Cambridge, Mass.: MIT Press, 1992. Reprinted with permission of MIT Press.

The Universal Reacher in Ronald Finke, Thomas Ward, and Steven Smith, *Creative Cognition: Theory, Research and Applications.* Cambridge, Mass.: MIT Press, 1992. Reprinted with permission of MIT Press.

An Imaginary Creature created in experiments on exemplar generation in Ronald Finke, Thomas Ward, and Steven Smith, *Creative Cognition: Theory, Research and Applications.* Cambridge, Mass.: MIT Press, 1992. Reprinted with permission of MIT Press.

An imaginary creature generated under the constraint that it had to possess feathers in Ronald Finke, Thomas Ward, and Steven Smith, *Creative Cognition: Theory, Research and Applications.* Cambridge, Mass.: MIT Press, 1992. Reprinted with permission of MIT Press.

Every effort has been made to contact the copyright holders of each of the selections. Rights holders of any selections not credited should contact W. W. Norton and Company, Inc., 500 Fifth Avenue, New York, NY 10110, in order for a correction to be made in the next reprinting of our work.

Index

Page numbers in *italics* refer to illustrations.

About the Author

David Perkins received his Ph.D. in mathematics and artificial intelligence from the Massachusetts Institute of Technology in 1970. Since 1971, David Perkins has served as codirector of Project Zero, a research group at Harvard Graduate School of Education concerned with learning, intelligence, creativity, and understanding in child and adult contexts. He has published and spoken widely on these subjects.